GEOMETRIC PATTERNS WITH CREATIVE CODING

CODING FOR THE ARTS

Selçuk Artut

Apress®

Geometric Patterns with Creative Coding: Coding for the Arts

Selçuk Artut
İstanbul, Türkiye

ISBN-13 (pbk): 978-1-4842-9388-1　　　　ISBN-13 (electronic): 978-1-4842-9389-8
https://doi.org/10.1007/978-1-4842-9389-8

Managing Director, Apress Media LLC: Welmoed Spahr
Acquisitions Editor: Shivangi Ramachandran
Development Editor: James Markham
Coordinating Editor: Jessica Vakili

Distributed to the book trade worldwide by Springer Science+Business Media New York, 1 NY Plaza, New York, NY 10004. Phone 1-800-SPRINGER, fax (201) 348-4505, e-mail orders-ny@ springer-sbm.com, or visit www.springeronline.com. Apress Media, LLC is a California LLC and the sole member (owner) is Springer Science + Business Media Finance Inc (SSBM Finance Inc). SSBM Finance Inc is a **Delaware** corporation.

For information on translations, please e-mail booktranslations@springernature.com; for reprint, paperback, or audio rights, please e-mail bookpermissions@springernature.com.

Apress titles may be purchased in bulk for academic, corporate, or promotional use. eBook versions and licenses are also available for most titles. For more information, reference our Print and eBook Bulk Sales web page at http://www.apress.com/bulk-sales.

Any source code or other supplementary material referenced by the author in this book is available to readers on the Github repository: https://github.com/Apress/Geometric-Patterns-with-Creative-Coding. For more detailed information, please visit http://www.apress.com/source-code.

Printed on acid-free paper

For Kaya, the best gift life has given me

Contents

About the Author

Selçuk Artut's artistic research and production focus on theoretical and practical dimensions of human-technology relations.

Artut's artworks have been exhibited at AKM (Istanbul 2022), Akbank Sanat (Istanbul 2022), Dystopie Sound Art Festival (Istanbul 2021 and 2019, Berlin 2018), Zilberman Gallery (Berlin 2019), Moving Image NY (New York 2015), Art13 London (London 2013), ICA London (London 2012), Art Hong Kong (Hong Kong 2011), and Istanbul Biennial (Istanbul 2007) and received coverage at Artsy, Creative Applications, CoDesign, Visual Complexity, and CNN GO. He holds a PhD in Media and Communications from European Graduate School, Switzerland.

An author of seven books and an editor of one, Artut is a professor at the Visual Arts and Visual Communication Design Program, Sabanci University, Istanbul. He has been releasing several albums as a member of a post-rock music band Replikas since 1998. In 2016, Artut cofounded an audiovisual performance duo named RAW (www.rawlivecoding.com), which produces works through creative coding and live coding techniques.

More info about Selçuk Artut can be found at www.selcukartut.com and at @selcukartut (Instagram).

About the Technical Reviewer

Alp Tuğan is an adjunct faculty member in the Department of Communication Design at Ozyegin University with interests in sociotechnology, creative coding, generative art, and sound. His generative audiovisual works have been included in several exhibits and events. His articles on sound technology were published in *Volume Magazine* between 2006 and 2009. Tugan is also a cofounding member of the live coding duo RAW. More information about his work can be found at www.alptugan.com/.

Acknowledgments

I am grateful to the following people for their helpful commitments on various dimensions: Alp Tuğan, for being the technical reviewer; David Wade, for being so kind to write a foreword; Serap Ekizler Sönmez and Ezgi Kara, for elaborating on the geometric patterns' structure; the team of evaluators, Ebru Bici Nasır, Deren Ertaş, Can Büyükberber, and Bahar Türkay; my students in my Creative Coding class, for evoking me to explore new ways; Murat Durusoy for the profile photo; and Apress friends, for making this dream come true.

Foreword

David Wade[1]

In recent times, there has been a welcome degree of recognition of the enormous contribution made by early Islamic scientists and mathematicians to world knowledge; this is long overdue. In particular, the role of Islamic learning in the recovery and development of Classical texts, and the subsequent transmission of this huge body of scholarship to Medieval Europe, is now widely acknowledged. It is no longer believed that there was an unbroken chain of learning in Europe from the Classical period to the early Modern – and it is now properly seen that the great rekindling of interest in all the sciences that occurred during the Renaissance was largely fueled by Islamic erudition (Montgomery Watt, 1972; Saliba, 2011).

The "Golden Age" of Islamic science flourished in the early Abbasid period (in Baghdad between the eighth and tenth centuries CE) and produced such outstanding scholars as al-Khwarizmi, al-Kindi, and Omar Khayyam. But Islamic science, building on the foundations of Classical, Parthian, and Indian knowledge, was to continue making important advances in various centers around the vast Islamic dominions for some centuries to come. Wherever the social, intellectual, and economic conditions were conducive, advances were made – particularly in such fields as mathematics, astronomy, optics, and medicine. Nevertheless, science, whose pagan philosophical associations were never entirely forgotten, continued to be regarded with some suspicion by the orthodox. Scientifically minded thinkers under royal patronage were usually afforded a measure of protection from the more zealous religious critics – who were in any case less concerned with such abstract fields of study as mathematics. It can be said then that, at least in higher intellectual levels, the study of mathematics (particularly of geometry) was well established and widely taught, throughout the Medieval Islamic world, and that the Alexandrian Platonists, Euclid and Ptolemy, retained their positions as the revered progenitors of geometry and astronomy, respectively.

---

[1] David Wade is an artist and architect, as well as the author of several books: *Pattern in Islamic Art*, 1976; *Geometric Patterns and Borders*, 1982; *Crystal and Dragon: The Cosmic Dance of Symmetry and Chaos in Nature, Art and Consciousness*, 1993; *Li: Dynamic Form in Nature*, 2003; and *Symmetry: The Ordering Principle*, 2006.

This text is compiled from various articles published by David Wade on https://patterninislamicart.com/.

The enormous achievement of the Golden Age in collecting, translating, disseminating, and building on the philosophies and sciences of the Classical past was of immeasurable importance. Ironically, the Western infidel nations were eventually to be primary beneficiaries of this fund of knowledge; it was instrumental in pulling Europe out of its Dark Ages and laid the foundations of the Enlightenment. The names of Islamic scholars feature on the first pages of most histories of European science, mathematics, medicine, and astronomy – and of course the rich tradition of Classical philosophy was first conveyed to the West in Latin translations from Arabic sources.

But the cultural attainments of the Hellenized civilization that the early Muslims encountered, and the rich traditions of knowledge that their successors so readily adopted, did have a lasting effect on Islamic civilization; in fact it helped to create it. Many of the norms of Islamic life were formed as a result of this contact. In essence there was a continuity of late-Hellenistic cultural values into the Islamic sphere. The clearest outward expression of this legacy may be found in Islamic art.

Islamic art has a recognizable aesthetic signature that somehow manages to express itself across an entire range of productions. The "language" of this art, once established, was readily assimilated by each of the different nations and ethnicities that were brought within the Islamic sphere. Assimilated and built upon, because every region, at every period, produced its own versions of this supernational style. Much of the art of Islam, whether in architecture, ceramics, textiles, or books, is the art of decoration – which is to say, of transformation. The aim, however, is never merely to ornament, but rather to transfigure. Essentially, this is a reflection of the Islamic preoccupation with the transitory nature of being. Substantial structures and objects are made to appear less substantial; materials are de-materialized.

From a purely doctrinal viewpoint, geometrical designs, being free of any symbolic meaning (which is the case in Islamic art), could convey a general aura of spirituality without offending religious sensibilities. In addition, the purity and orderliness of patterns and symmetries could evoke a sense of transcendent beauty, which, at best, would free and stimulate the intellect (rather than trap it in the illusions of mere representation).

The widespread use of complex patterning in the Islamic world, over many centuries, clearly indicates that this mode of expression satisfied something integral to the Islamic ethos – not least because the use of the geometric mode is not expressed in any single unbroken tradition. Technically speaking, there are any number of ways to create high levels of complexity in Islamic geometric patterns, and in the end, it is only the broad geometric theme itself that is constant. That is to say that although there is tendency toward greater geometric complexity through time in the various Islamic regions, this is expressed in a wide range of different styles and approaches. It is however intriguing that these humble Muslim craftsmen, over time, managed to uncover and exploit for decorative purposes many of the possibilities allowed of formal symmetrical arrangement.

There is little doubt that artists/craftsmen traveled (voluntarily or otherwise) across widely separated regions and that at different times and places pattern books and working drawings would have facilitated transmission. Given Islam's fractured history, it is likely that traditions were broken, sometimes to be resumed in subtly different ways, and that patterns would have been adapted from one medium to be used in another. Taken as a whole, this long tradition indicates that over time, experimentation vied with adherence to established forms. There is a thread of continuity in the use of decorative ornament in Islamic Art, but the real constant was a thorough appreciation of the underlying geometry of plane division and a mastery of what can only be described as artistic geometry. This involved a keen awareness of an unspoken set of rules, which involved a strong sense of symmetry and a preference for the careful, balanced distribution of elements within framing panels. In the end, it is these criteria, rather than any mathematical formulae, that go a long way to make Islamic art, in all its diversity, so recognizable, coherent, and distinctive.

Introduction

In this book, I intend to present multiple layers of issues, all of which are important to me. To begin, I would like to emphasize that our perspective on the history of art should be handled in an all-encompassing manner and that it should not belong to any particular group or period. I believe that the history of art should be understood as a complex dialogue between various periods and cultures and that this dialogue should be embraced for its richness and diversity. By exploring the works of artists, writers, and scholars from across the world and throughout the ages, we can gain an appreciation for the interdependence of art, culture, and history. Second, I would like to reveal that the abstractions that can be achieved through the art of geometry in retrospect are the source of inspiration for the media art of today. I want to bring to light the connections that have been hidden for so long between the geometric art of the past and the generative art of the present. By doing so, I hope to highlight the power of geometry and abstract art to influence and shape our current artistic practice. Third, I would like to encourage future research on this topic by proposing as a definition the dialectical creativity that emerges from the unification of humans and technology. Through this definition, I hope to demonstrate the potential of media art to synthesize different disciplines into a meaningful whole.

This book presents a series of workflows for analyzing twenty three distinct geometric patterns and producing them through creative coding. In the "Patterns Index" section that follows, you will see that I have ordered these patterns in a list from easiest to most difficult to handle. Working through the book, I hope that you will get a solid understanding of the computer-generated geometric patterns, along with code samples, to create your own unique designs. At the beginning of each workflow, you will also find a piece of artwork that was created using that particular motif. Utilizing the open source code resources that are at your disposal will also allow you to produce works of this nature. My goal is to maintain the traditional anonymous stance on the art of geometry through the use of open source collaboration. Sharing is caring!

During the time that I was writing the book, I was filled with a lot of excitement and pleasure, and I hope that you can feel the same way. I also hope that you can take away from the book an appreciation for geometry and abstract art and their influence on our culture.

Patterns Index

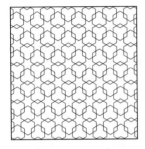

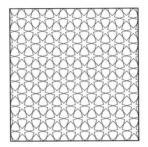

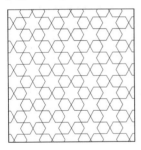

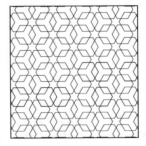

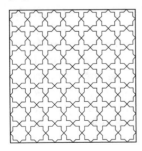

Workflow #07

Workflow #08

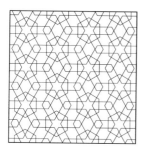

Workflow #09

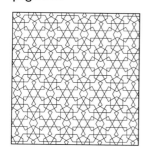

Workflow #10

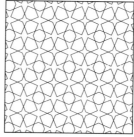

Workflow #11

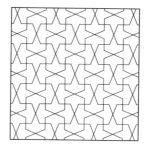

Workflow #12

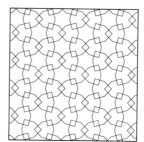

Workflow #13

page 222

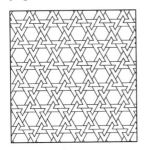

Workflow #16

page 260

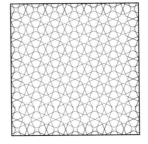

Workflow #14

page 234

Workflow #17

page 272

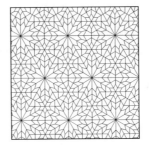

Workflow #15

page 246

Workflow #18

page 292

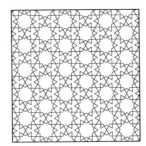

Workflow #19

page 302

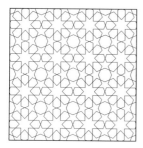

Workflow #20

page 314

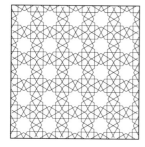

Workflow #21

page 328

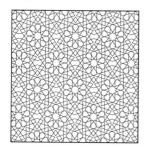

Workflow #22

page 352

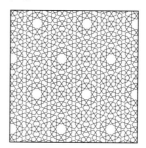

Workflow #23

page 386

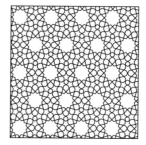

The Fascinating Universe of Geometric Patterns

Geometric patterns are everywhere in our everyday lives, from the tiles on the floor to the wallpaper on our walls and even the clothes we wear. It is a type of artistic expression that dates back hundreds of years. Some geometric patterns are the product of human ingenuity, while others arise spontaneously in the environment. If you look around, you can find many examples of geometric patterns in nature, each with its unique combination of dimensions, colors, and shapes. Patterns, in general, have a certain regularity within themselves and have a distinctively repetitive visual structure. As a result, geometric patterns are a common form of visual expression found worldwide, both in natural and manufactured environments. In geometry and mathematics, patterns are the building blocks of the physical world. On the other hand,

© Selçuk Artut 2023
S. Artut, *Geometric Patterns with Creative Coding*,
https://doi.org/10.1007/978-1-4842-9389-8_1

geometric patterns are a specific framework that uses the science of geometry to explore alluring dimensions in building abstract forms. It is assumed that there is one way to perceive the complexity and beauty of such patterns, and that way is mathematical.

Known as the science that deals with spatial relationships, geometry derives from the ancient Greek terms *geo*, which means "earth," and *metron*, which means "measurement." It studies shapes, sizes, and relationships and is helpful in many fields, such as architecture, engineering, and the sciences. You can construct geometric figures by initiating with angles; the size you give each building element will determine the type of figure you can create. Once the angles are established, you can construct shapes by adding lines and circles of different sizes and lengths. Geometric patterns provide an understanding of the basic laws that govern shapes and allow for a greater appreciation of beauty and complexity.

Geometry in the early days was a collection of empirically discovered observations based on spatial problems in daily life, such as calculating areas of land. Among many others, the Egyptians and the Babylonians are considered the frontiers of the science of geometry. These civilizations could calculate the area of a circle with great dexterity. Concerning the Plimpton 322 tablet, even it is claimed that the Babylonians knew the Pythagorean theorem 1500 years before the Greeks. Nevertheless, it was in the Greeks that the science of geometry gained groundbreaking developments.

Geometry was considered the pinnacle of mathematics by the ancient Greeks because they could find accurate inferences about their observations of the world. According to the philosopher Plato, geometry was key to unlocking the universe's secrets. Nevertheless, Plato also realized that while mathematics was powerful enough to make precise calculations, observations would never quite match up in real life. With all the skepticism, he proposed the idea of the Theory of Forms that dismisses the material world as an imitation of reality. This theory implies that there are two separate worlds: one is abstract, and the other is perceptible. No matter how finely detailed a mathematical phenomenon was, it was inevitable that it would only be valid in an abstract sense. Many scientists following the same tradition were keen to use the language of mathematics to explore ambiguous areas of science without the intention of having direct real-life consequences. Despite all the uncertainty, one mathematician named Euclid from ancient Alexandria, around 300 BC, wrote a mathematical treatise consisting of 13 books called Elements, a compilation of definitions, postulates, theorems, constructions, and proofs of the mathematical theorems. This treatise was so influential in mathematics that it significantly strengthened the system of axiomatic thinking in solving geometric problems. In this series of books, Euclid listed a set of axioms to provide a solid foundation for geometry problems that could be handled using a simple compass and ruler.

In many societies from ancient times to the present day, the use of geometric elements in works of art has been at the center of the attention of artists. According to Wichmann and Wade,[1] geometric designs have long attracted the attention of Muslim designers and artisans. They have a spiritual or otherworldly aura without being associated with specific doctrinal propositions. Likewise, they avoid carrying any symbolic value in Islamic content. First, they allow artisans to showcase their expertise and the complexity of their craft and simultaneously amaze and fascinate with their pure complexity. Not to be overlooked is the fact that the development of the art of geometry within the context of Islam was not a mere coincidence, given the popularity of mathematics.

When, in the sixth century, Islam emerged in the Arabian Peninsula, which is geographically defined as the Near East, pagan belief was quite common in the region. There were many idols representing the belief of paganism in the Kaaba in Mecca, where Prophet Muhammad was born. Nonetheless, the message that Prophet Muhammad received in the revelations was a warning that people can attain salvation in the hereafter through devotion to the one and only God.[2] Being a monotheistic religion, Islam avoided all kinds of idol worship. Following his appointment as a religious leader by the people of Mecca, Muhammad's first act was to remove the 360 idols surrounding the Kaaba. Mecca was previously a major pagan pilgrimage site. The fact that Islam forbade the worship of idols was already a practice that existed in earlier monotheistic religions. Unlike what was practiced in the past, the religion of Islam did not see the need for any intercession between God and the individual. It has always been thought that the relationship between man and the divine God must occur with a degree of purity, independence, and transparency. Considering this aspect, the place of the art of geometry in the religion of Islam is better understood. Islam assumes that any figurative image presented in places of worship can be perceived as iconoclastic and thinks such images will interfere with the direct devotion between the individual and God. For this reason, since geometric patterns do not have a dogmatic structure that will create a directed judgment in the individual, they have been much more accepted in Islamic art with the abstract content they offer.

Another reason for the widespread use of geometry in Islamic arts is the intense interest in science and technology, which gained momentum, especially in the Middle East, in the eighth and ninth centuries. In this Islamic Enlightenment period, also known as the Abbasid period, Classical Greek and Roman teachings were blended with Persian and Hindu teachings in Baghdad, which was by then the capital of the Abbasids. Baghdad lived its heyday during the reign of Caliph Harun al-Rashid. During this period, ancient Greek texts

[1] Wichmann, B., & Wade, D. *Islamic Design: A Mathematical Approach*. Cham, Switzerland: Birkhäuser (2017).
[2] Esposito, E. *The Oxford History of Islam*. Oxford University Press (1999).

were extensively translated into Arabic. During the authority of the Abbasids, the "House of Wisdom" (Bayt al-Hikmah) was established as a leading intellectual center. This center was a crucial component of the Islamic Golden Age that developed with the ongoing translation movement. Over time, there have been very eminent scientists who have grown up in the House of Wisdom. For example, al-Khwarizmi made significant contributions to the science of mathematics with his studies in the field of algebra. The emergence of the word *algorithm* is also based on the work of al-Khwarizmi. The results of Ibn al-Haytham's using geometry extensively made significant contributions to optics and visual perception principles. With the dominance of mathematics and geometry in many areas of the Islamic world, the art of geometry has also undergone structurally significant developments.

The most distinctive feature of Islamic geometry art is the symmetrical and proportional arrangements between the shapes. This art has a technical structure that enables highly complex designs to be obtained with a simple compass and ruler and has explicit content consisting of systematic methods in which the delicate balances between shapes and the resulting order must be carefully constructed. Unfortunately, we have very little information and documents about how this art was performed in its time. Topkapi, Tashkent, and Mirza Akbar Scrolls, among the documents in hand, are important sources for understanding Islamic geometric ornament works. The Topkapi parchment was created by master builders in the late Medieval Iranian world and contained a rich repertoire of geometric drawings for wall surfaces and vaults.

Today, various methods and techniques are applied to draw Islamic geometric patterns. Generally, the initial geometric shape positioned at the center of the drawing acts as the main determining factor in shaping the geometric structure that will emerge as a result. The compass and straightedge were the only tools available for generations to create polygons and the necessary angles. As a result, all these polygons are based on templates made from grids of circles and come from harmonic subdivisions of circles.

For this reason, one of the most common methods is to generate these essential geometric determinants using a compass and a ruler by applying a few procedural operations. Apart from the drawing method based on the mentioned procedures, some ways deal with the subject mathematically. This book presents a systematic approach to creating a geometric pattern by calculating the necessary reference points of a motif, which is the main constructive component of a tessellated image. Through this approach, the underlying mathematical principles can be studied and explored to understand a pattern's structure and its possibilities for creative expression. The main focus of this book is to provide insight into how certain geometric shapes can be used to form an overall pattern, as well as how these shapes interact with one another and create a unified composition.

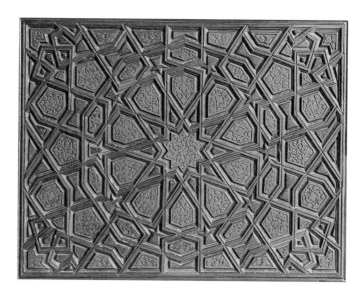

Geometric pattern from the Yavuz Sultan Selim Mosque, Istanbul, Turkiye

Geometric Pattern from the Shah-i Zinda, Samarkand, Uzbekistan

Geometric Pattern from a wooden door in Khiva, Uzbekistan

Algorithmic Structures in Art

Algorithms

Throughout history, there has always been a bilateral interaction between art and technology at different levels of involvement. Technology has expanded the creative expression of artists by developing new mediums, tools, and subjects at every stage. Although it seems that technology has been sailing away from the discipline of arts, indeed, the word "technology" is derived from the Greek word "technikon," which is connected to the word "techne." Furthermore, techne, the epistemic root of the word "technology," can mean either craft or art. Art and technology were thus seen to be inseparably linked prior to antiquity. This shows that technology and art have a deeper connection than is sometimes thought. While these two concepts are often seen as distinct and opposed, art has a long history of being intertwined with technology.

Nevertheless, the impact of technology on art is inevitable. For example, many painters in the past and still today use a variety of distinctive recipes to discover novel colors that inspire their paintings. If we take a closer look at Impressionism, we can see that a change in the use of brush types severely impacted the aesthetics of their paintings. Historically, before Impressionism,

© Selçuk Artut 2023
S. Artut, *Geometric Patterns with Creative Coding*,
https://doi.org/10.1007/978-1-4842-9389-8_2

weasel hair was mainly used to make brushes. However, in Impressionism, landscape painters intentionally preferred pig hair because it was thicker and more rigid. As a result, this deliberate move in technological preference has a crucial role in establishing a new artistic look.

If we explore how this advance in technology has been achieved, we can see that one of the main reasons is that, as humans, we have the advantage of making our ideas a reality by consistently applying a set of rules and procedures and then passing this knowledge on to future generations. Consequently, repetition and order are considered essential elements of our structural thinking. To consistently distinguish a repetitive phenomenon from a disorder and make it comprehensible, it is necessary to examine the building elements that form a pattern and thus cause such consistency in the repetitive process. These patterns we perceive through empirical observation or theoretical analysis contribute to advancing the civilization we live in today. To be able to transfer and convey the methods of recreating these repetitive patterns to others, people needed to develop a descriptive scheme for converting these gained experiences into a systematic set of rules.

In computer science, we define an algorithm as a finite sequence of explicit instructions employed to solve specific problems using a computational system. The word "algorithm" comes from the name of the Persian mathematician Muhammad ibn Musa al-Khwarizmi, who lived in the ninth century. Al-Khwarizmi was a polymath from a region called "Khwarazm," and that's where the root of the word has been derived from. His book titled *The Compendious Book on Calculation by Completion and Balancing* (*al-Kitāb al-Mukhtaṣar fī Ḥisāb al-Jabr wal-Muqābalah*), which popularized algebra in Europe, was significant in providing systematic approaches to solving linear and quadratic equations. The book is considered the foundation text of algebra, establishing it as a distinctive field of study through the compilation of general principles for solving quadratic equations.

Today, algorithms have become widely used in our everyday life. Computers rely heavily on applying algorithms to compile a set of functions to produce effective results. Algorithms are the connecting bridge between our logical thinking and computational operations. For example, when we consider a simple number sorting problem, although we can do this intuitively when we have a small set of numbers, it becomes evident that an algorithmic solution is required as the set of numbers increases.

Let's assume we have a set of numbers as follows:

A = {40, 12, 18, 33}

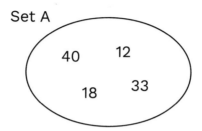

By looking at the overall picture of the set of numbers and identifying them, we can intuitively sort them as 12,18,33,40 in ascending order very quickly.

However, this time let's assume that we have a larger set of numbers as follows:

B = {24, 27, 44, 31, 19, 15, 21, 45, 8}

Now it becomes a tedious task to order them by just looking at the overall picture of the set of numbers.

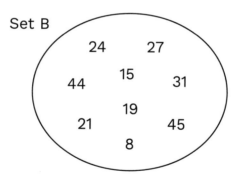

Various sorting algorithms can be applied in computer science, such as insertion sort, merge sort, quick sort, bubble sort, distribution sort, etc.

As an example, let's see how the bubble sort algorithm would sort this list of numbers in an ascending order.

Bubble Sort Algorithm

The algorithm begins at the start of the list. The first two elements are compared, and if the first is greater than the second, they are switched. Up to the finish of the list, it does this for every pair of neighboring elements. After then, it repeats the process with the first two elements until the final run has produced no swaps.

Pseudocode:

```
do
    swapped = false
    for i = 1 to indexOfLastUnsortedElement-1
        if leftElement > rightElement
            swap(leftElement, rightElement)
            swapped = true; ++swapCounter
while swapped
```

Iterations:

Iteration 1: {24, 27, 31, 19, 15, 21, 44, 8, 45}

Iteration 2: {24, 27, 19, 15, 21, 31, 8, 44, 45}

Iteration 3: {24, 19, 15, 21, 27, 8, 31, 44, 45}

Iteration 4: {19, 15, 21, 24, 8, 27, 31, 44, 45}

Iteration 5: {15, 19, 21, 8, 24, 27, 31, 44, 45}

Iteration 6: {15, 19, 8, 21, 24, 27, 31, 44, 45}

Iteration 7: {15, 8, 19, 21, 24, 27, 31, 44, 45}

Iteration 8: {8, 15, 19, 21, 24, 27, 31, 44, 45}

Iteration 9: {8, 15, 19, 21, 24, 27, 31, 44, 45}

The algorithm stops here.

Even though this long-lasting set of operations is not a very efficient task for us, it is relatively easy for computers to accomplish it with some repetitive iterations. As a result, the only thing that really matters is optimizing the complexity of the algorithms in order to reduce the amount of computational expense they require.

Algorithms offer various solutions to the usual problems of our daily life. Sometimes they help us make a trip in the shortest way, and sometimes they allow us to cook a meal in the most delicious way. Algorithms, with their functionality extending even to our social relations, have become an indispensable part of the world in which we live in constant progress. Interestingly, as a concrete example, the stable marriage problem, which focuses on the relationships between couples and their impact on a society's well-being, deals with finding a stable match between two equally sized item groups with ordered preferences. It was first stated in a paper published in 1962 by Gale and Shapley. Although its name seems to be just about addressing a marriage problem, the algorithm is indeed trying to find a solution to the optimum matching problem for the elements of two separate sets. For this reason, there is a wide range of applications within a framework ranging from university admissions to hospital residents' problems in the real world.

The stable marriage problem states that

For two sets of an equal number of men and women, there exists an optimal matching given that

1. Participants rate members of the opposite sex.

2. Each man lists women in order of preference from best to worst.

3. Each woman lists men in order of preference from best to worst.

It assumes that optimal matching happens when everyone is matched monogamously with no unstable pairs. In a matching process, an unmatched pair is regarded as unstable if the man and woman prefer each other to their current partners. Suppose, without any discrimination, that the set of men is the set of applicants who propose marriage, and we are looking at a man's optimal result. The aim is to loop through all men until no men are single. Each unmatched man makes an offer to his most preferred woman. He checks if the woman he proposed to is not matched; if appropriate, they are both engaged. If the woman is already matched, either she says no to him or she breaks a once-made engagement if she meets a better option.

Let us see how it works with an example. Assume we have associated each person with an ordered preference list containing all the members of the opposite sex.

In the first iteration, Kaya proposes to Gizem who is on the top of his preferred list. Gizem is unmatched, so they are engaged.

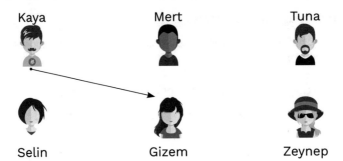

In the second iteration, Mert proposes to Selin who is on the top of his preferred list. Selin is unmatched, so they are engaged.

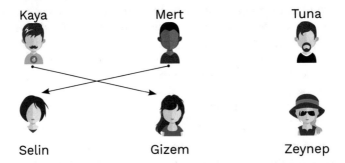

In the third iteration, Tuna proposes to Selin who is on the top of his preferred list. Selin is already matched, but since she prefers Tuna more, she breaks the engagement and matches with Tuna.

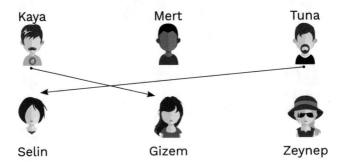

In the fourth iteration, Mert proposes to Selin, but she is already matched with Tuna and they are happy. So he offers to Gizem, but she is engaged with an optimum choice too. So he offers Zeynep, and Zeynep accepts. Since everyone now has a match, the algorithm quits.

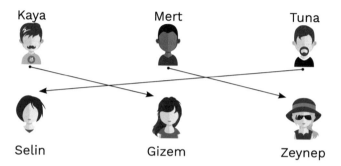

This problem, which Gale and Shapley presented in 1962, has been investigated by many scholars, and different solutions have been addressed to it today. The algorithm has even been implemented on platforms gaining popularity in network-based applications such as matchmaking. The use of algorithms continues to provide useful solutions for the problems we face in our daily lives. From computing and organizing our data to finding us the best routes to take during our commutes, algorithms have become essential in ensuring efficiency and accuracy in our everyday activities.

Automata

Algorithms are basic procedures that use a set of rules to perform a specific task. They are also an important building block of autonomous systems due to the consistent and repetitive execution of the recorded procedures. Likewise, we refer to automata as self-operating machines or control mechanisms created to automatically carry out a specified sequence of tasks or react to preordained instructions. A classic example of an automata is the player piano, which plays its music without human intervention. Automata differ from algorithms in that they are physical machines with the capacity to carry out programmed tasks and even make decisions based on their environment. Automata have been around for centuries, but they have become increasingly popular in the modern world due to their potential to reduce human labor and increase efficiency.

Automata have unprecedented levels of advancements in the employment of technology. For example, automata built around the eighth century in the Arabian Peninsula used waterpower to embody complex mechanical systems to perform exciting tasks. In the ninth century, the Banu Musa Brothers wrote *The Book of Ingenious Devices*, consisting of the extraordinary inventions they devised. This book shows a self-powered hydroelectric mechanical organ working with a revolving cylinder that holds a recorded set of actions. As an Arabic polymath, Al-Jazari wrote about complicated, programmable humanoid automata in *The Book of Knowledge of Ingenious Mechanical Devices* in 1206. He developed highly advanced mechanical systems that would automatically play music for you using hydropower during a sailing trip.

Likewise, the automata built in Europe during the Renaissance used mechanical cogs and wheels to simulate various life forms. For instance, in 1737, Jacques de Vaucanson designed an artificial duck digesting grain and even defecating it. While the watchmakers of the eighteenth century tried to demonstrate their superior skills by creating sophisticated automatic machines with fascinating features, the automata of the time were also masterfully designed and produced with meticulous attention to minor details.

One of the most striking examples of European automata is the trio set of a musician, a writer, and a draftsman produced by the famous watchmaker Jaquet-Droz family. These three humanoid automata from the eighteenth century are precision marvels that exhibit great technological competence for the time. Created by putting together approximately 6000 pieces, the writer automata has a very advanced level of sophistication compared with the other two automata in the set, as it can be programmed for up to 40 letters or signs, as well as being able to perform a certain number of predefined movements flawlessly over and over again. The common feature of all automata was that the pattern they had to repeat was carved on a physical surface in the shape of a cam structure. Therefore, in an automata system, the mechanical construction can read the recorded information on a cam surface and follow a series of events based on that information to accomplish a specific task. While these eighteenth-century automata were cutting-edge for their time, it is remarkable to consider how much more advanced our technology has become since then.

Technology's Impact on Art

Algorithmic approaches and inferences based on discovering novel patterns, examples of which we can see throughout history, have also formed the basis of the mechanization of today's human civilization, accompanied by the soul of technology. With the spread of steam power and the invention of machine tools, the Industrial Revolution inevitably impacted modern life. Throughout history, there has always been a two-way interaction between art and technology at various levels of participation. It is not possible to consider the evolution of art independently of the social developments it is in. Accordingly, art has been mainly influenced directly by the environment in which society is, and the content of art continues to shape with the changes in society. Art, in its marching augmentation, has brought about a multitude of artistic movements with distinctive characteristics that emerge from the sensational influences of various times in history. These movements have generally been influencing each other with transitional boundaries. Occasionally distinct artistic movements form over time as a reaction or a derivation of a certain style and conception. The two very distinctive though short-lived movements of the early twentieth century known as the Futurism and Dadaism Art movements played an existential role in building bridges between art and technology.

Futurism art has been depicted throughout history as an art movement profoundly influenced by the world's revolutionary political and technological developments. In 1909, Futurism art was acquainted for the first time with large audiences through Filippo Tomaso Marinetti's article "The Founding and Manifesto of Futurism," which was published on the front cover of the newspaper *Le Figaro* in Paris. With the invention of new imaging technologies such as chronophotography, Futurist artists studied physical exertion as a means to depict the fragmentation of movement in still images. When a continuous action was broken into its constituent elements, it was observed that continuity could also be described as a kinesis of sequential iterations. Furthermore, advanced use of machinery allowed sophisticated tasks to be analyzed in detail to construct its building components to formulate a fabrication system. Consequently, Futurist artists were heavily inspired by the flourishing technological advancements in the early twentieth century and portrayed these inspirations in the various forms of art.

The charming elevation of the power of machines profoundly dominated Futurist artists. Consequently, Futurism ideology aims to convey the dynamism of the modern world with an understanding deeply influenced by the rapid advancements of the era of science and technology. When their avant-garde approach to the arts was flavored with the praising of technology, their works of art primarily reflected the range of notions such as speed, power, movement, temporality, electricity, cityscape, mechanical modes of production, modernity, etc.

Futurist painters have tried to break down a visual experience into its abstract elements. With the influence of chronophotography on Futurist painters, the notion of transforming movements into a sequence of silhouettes became a subject of interest for the composition of paintings. When the Futurist painter Giacomo Balla first saw Etienne-Jules Marey's chronophotography works in 1900, he was captivated by the framing of suspended kinesis in the appearance of stasis. Balla's approximation in depicting a movement on a painting was celebrated in his famous painting "Dynamism of a Dog on a Leash ("Dinamismo di un cane al guinzaglio") in 1912. In this painting, the rapid movements of the dog's feet and the lady's walk cycle were superimposed to capture their motion in a single moment. The suspended moment of temporality and dynamic equilibrium has contributed to the feeling of the painting with an exalted feeling of desire to elongate in a fourth dimension.

The use of repetitive forms on canvas has been frequently implemented in Futurist painters' compositions. In Giacomo Balla's works based on the themes of speed and motion, his painting style is characterized by extensive use of repetition techniques. In his painting titled "Lines of Movement and Dynamic Succession" (1913), the layering of geometric shapes and curvature representation of sequential copies of visual elements on superimposed juxtapositions imply his vigorous attempt to portray the infiltration of the movement into imagery. Placing a flying bird's repetitive appearance in a consecutive structure enhances the idea of illustrating a continuous movement as a breakdown of precise discrete samples.

Balla's articulation of the same figurative approach may be observed mainly in his other paintings, including "Speed of a Motorcycle" (1913), "Abstract Speed" (1913), and "Flight of the Swallows" (1913). In a famous classic example of Futurist paintings, "Speed of a Motorcycle," the repeated shell image and range of curvatures designate an abstract shape to signify an accelerated motorcycle's displacement. While Balla attracts the viewers thoroughly into his radiant composition, he adopts the technique of using a sequential arrangement of the same geometric shape that has been recursively rotated from its center of origin.

In this painting, we observe a repetition of a single object and a recursive manipulation of its replicas with variance in its positioning, rotation, and scale properties. Similarly, handling the control of derivations in the building blocks of a visual material has been a widely visited point of departure for the computer-based generative arts. In many Futurist artworks, we observe that painters and sculptors essentially try to capture the object in action, movement and speed. As for the art style, it is possible to say that dynamism is one of the main features of Futurism. Although Futurism did not want to stay within the confines of a particular visual style, many works in Futurism included common visual elements such as rigid, geometric lines and planes that were similar to the cubist style.

Similar to Futurism in that it sought to change art history, the Dada movement, which first appeared in Zurich in 1919, had a different focus on the content. Dadaism was an art movement that opposed the brutality of conflict during the First World War. It began as a reaction to the atmosphere of war while also trying to eliminate the art and artistic aesthetic of the 1920s. Less organized than Futurism, they supported a rebellion against artistic modernism, a movement that strived for the celebration of art over everything else. This challenged their belief that art was supposed to be the central work of culture. Art was seen as a sort of reflection of humanity and as a move toward a higher form of humanity. While the aesthetic of Futurism was modern and dynamic, the aesthetic of Dada was more like everyday movements and social issues that were being developed in the 1920s. According to the Dadaists, art was supposed to make people feel good, and since the war had made them feel terrible, artists had to fight against that to create new art.

The popularity of Dada increased as it spread across Europe and manifested itself in a wide variety of forms. Through such figures as Jean Arp, Hugo Ball, Marcel Duchamp, George Grosz, Hannah Höch, Francis Picabia, Hans Richter, Kurt Schwitters, and Tristan Tzara, Dada put itself at the forefront of the new art movements. In addition to the methods Dada used to create unconventional works of art, it also had a humorous side. They used humor to defuse the absurdism of modernist art.

With little concern for producing well-founded and meaningful works of art, they preferred to put randomness, spontaneous creativity, and stirring emotional responses first. For example, Tristan Tzara, a poet and performance artist, was describing his methods of writing poetry in which he used chance operations with a specific set of instructions:

Take a newspaper.

Take some scissors.

Choose from this paper an article of the length you want to make your poem.

Cut out the article.

Next carefully cut out each of the words that makes up this article and put them all in a bag.

Shake gently.

Next take out each cutting one after the other.

Copy conscientiously in the order in which they left the bag.

The poem will resemble you.

And there you are – an infinitely original author of charming sensibility, even though unappreciated by the vulgar herd.

As it is observed in the preceding example, the emergence of incoherent results while using randomness was used as a creative factor in the fabrication of Dadaist art. The Dadaists relied on their distinctive techniques and instructions to create images, poems, and other works. These expressions and artistic directives can be interpreted as describing the use of algorithms to direct an artist to develop novel art forms spontaneously. Dadaism profoundly influenced many avant-garde art movements, such as Surrealism, pop art, and Fluxus, that came after it. Randomness is one of the most commonly encountered elements in contemporary artistic uses of code. The disconnect that exists between the artist's intentions and the finished product is one of the most important stimulants for original thought when it comes to code art.

Generative Art

Interdisciplinary hybrid approaches, which we encounter more frequently in the works of art produced in the twentieth century, have provided the fusion of many multimodal elements in one common structure. In the shadow of these tendencies, due to the widespread use of computers and the abundance of visualization systems in the 2000s, software-based generative art, with its rigorous effort to praise audiovisual content, has become an essential focal point in contemporary media art. The variety of functions that can be used on media artifacts to create a dynamic audiovisual composition is a major factor in advancing the generative arts. Following the examples of generative art that already took place a few decades ago, software-based generative art has advanced to a new stage of generative design and art production, with this latter term referring to the incorporation of software elements into media or art objects or methods.

The phenomenon of "generative art" enables creators to express themselves dynamically in the abstraction domain resulting from a collection of algorithms they intuitively anticipate. Although the sole authority to predict the final representation of a work of art lies with the artist, as a result of external factors such as randomness, user participation, data processing, etc., the work of art ultimately emerges from the momentum between human and technological interaction.

Generative art is an art genre that reflects the process as a representation. Traditionally, a painter or artist uses some form of material to create art, whether it be paint, pencil, clay, etc. In generative art, the artist uses elements such as abstraction, randomness, intuition, and procedures to create unique

S. Artut, *Geometric Patterns with Creative Coding*,
https://doi.org/10.1007/978-1-4842-9389-8_3

art forms. For this reason, generative art should be thought of as an action or gesture rather than a final product or as a process that initiates its own continued existence. Whereas new art movements are usually triggered by dissatisfaction with what has been done in the past, generative art is not generally designed to remedy specific problems in existing artistic forms. Generative art is a formal approach that has been around for a long time.

Despite a few examples of computer art that are considered "generative art," generative art should not be confused with computational art. Yet generative art existed long before the invention of computers. Although it can be thought of as a combination of computer technology and traditional art, a computer is not necessary to create generative art. Generative art is a term used to describe procedurally created works of art. In this sense, its history dates back to ancient times, even to the history of carpet weaving. Another early example would be the use of randomization in musical compositions. A system for randomly generating music from options that had already been composed was known as Musikalisches Würfelspiel, or "musical dice game." The dice randomly selects short pieces of music that would be pieced together to form a musical composition. The Mozart composition from 1787, which consists of 176 one-bar pieces of music, is one of the best-known examples of dice music. While using dice to compose music is hardly a common activity nowadays, randomness continues to play an important role in some forms of music. For example, early-twentieth-century composer John Cage used random elements in the composition of his pieces. Instead of a composer consciously choosing which notes to use, Cage used methods like the I Ching to determine which notes he would use in a piece. This composition technique can be seen as a type of aleatoric music, which is "music in which the composer utilizes an element of chance in determining the sequence of events and/or sounds."

Being a seminal figure in conceptual arts, Sol Lewitt extensively used generative art techniques to create his intricate works of art in the 1960s. In his wall drawings, he outlines in great detail the steps that other people are supposed to take in order to complete his artwork. For example, in the piece titled "Wall Drawing #49," Sol Lewitt outlines the drawing as a diagram and provides written instructions for how it is to be completed:

> *A wall divided equally into fifteen equal parts, each with a different line direction and color, and all combinations. Red, Yellow, Blue, Black pencil*

In a more recent example, Tim Knowles's "Tree Drawings" series features the artist attaching pens to the tips of the branches of various trees. On the paper, free forms are drawn by pens that move in response to the wind. His art takes the form of a more voluminous practice in which he incorporates elements of chance into the creative process by making use of apparatus, mechanisms, or systems that are beyond his conscious jurisdiction.

In addition to the pieces of generative art discussed earlier that make use of chance operations, the well-known piece of artwork that was composed in 1969 by Alvin Lucier, titled "I Am Sitting in a Room," is an outstanding example of the use of iteration in the process. Lucier records himself reading a text into a tape recorder and then plays the tape back into the room to re-record his playback. The new recording is played back in the same room and recorded again, and the cycle continues. The size and shape of the room cause some frequencies to be amplified, while others are muted in the iterative recording process. As the room's resonant frequencies take over, the words become indecipherable.

In his analytical essay "Generative Art Theory," Philip Galanter[1] explains the term as follows:

> *Generative art refers to any art practice in which the artist cedes control to a system with functional autonomy that contributes to, or results in, a completed work of art. Systems may include natural language instructions, biological or chemical processes, computer programs, machines, self-organizing materials, mathematical operations, and other procedural inventions.*

In this definition, we observe that authority over the creative process is shared between an artist and an accompanying system's involvement. In a generative work, the artist (1) supplies essential materials, parameters, and other information that feed into an autonomous process that results in an artwork, and/or (2) brings about a self-organizing process where new iterative processes are created, and/or (3) allows for an accompanying system to function on its own without active intervention. However, despite all the efforts to find a single definition of generative art that everyone agrees on, generative artworks can be described by a variety of systems and methods, which brings up questions about its historicity.

The origins of "generative art," produced by artists with the assistance of computers and software, can be traced back to the middle of the twentieth century. During the 1950s, a significant number of artists and designers worked with mechanical devices and analog computers to develop processes for generating artwork. For example, Benjamin F. Laposky, an American graphic artist and mathematician, was a pioneering figure who used a cathode ray oscilloscope with sine wave generators and a variety of other electrical and electronic circuits to create abstract art in the year 1950. He referred to the resulting works as "electrical compositions." Later on, we might be able to observe the practice of modifying the electromagnetic flow of electrons in a television set in order to generate abstract forms in Nam June Paik's works from the 1980s.

---

[1] Galanter, Philip. 2008. "What Is Complexism? Generative Art and the Cultures of Science and the Humanities." Proceedings of the International Conference on Generative Art, Generative Design Lab, Milan Polytechnic, Milan, Italy.

Very few people had access to computers in the early 1960s because they were still in their infancy. Therefore, computer scientists and mathematicians were among the first to use computers innovatively. The first computer-based "Generative Art" exhibition opened on February 4, 1965, by Georg Nees, in the Institute of Philosophy and Theory of Science at the University of Stuttgart in Germany. The drawings appeared to have elements that were chosen at random; the precision of straight lines, in combination with seemingly random elements, enabled new forms of variation and evolution to emerge. In 1968, Nees finished his PhD thesis, "Generative Computergraphik," which served as a signpost for the emerging field of generative art and design. The exhibition of works by A. Michael Noll and Béla Julesz at the Howard Wise Gallery in New York City later that same year was the first of its kind to take place in the United States. The Cybernetic Serendipity exhibition, which took place in 1968, is widely regarded as the foundational example of computer art exhibitions due to the profound impact it had on a large number of computer art pioneers and the breadth of audience it reached. The exhibition covered a variety of generative art examples, including music-making software and hardware as one of its sections. Musical effects were created by the devices, and computer-generated noises were played back on tape machines.

After the considerable amount of interest that was demonstrated in the Cybernetic Serendipity exhibition, it became possible to view the opening of numerous exhibitions that bring together technology and art, as well as the emergence of numerous habitats that can make it possible for these two fields to coexist. As more information about technological systems is made available to the public, more artists are able to develop these devices for their own purposes and display them in a much larger variety of artistic contexts. Recently, with the proliferation of the Internet and digitization becoming a part of our daily lives, we have seen a rise in popularity of open source software platforms that enable the creation of works of art. Some platforms, such as Processing, OpenFrameworks, Cinder, p5.js, and VVVV, have become more popular, thanks to the active participation of communities supporting open source programming environments. Artists who favored using programming as a creative element started producing works freely in open source environments and sharing these works with large audiences over the Internet. These software environments have had a profound impact on the widespread acceptance of the next generation of generative artworks that blend computer technology with abstraction, nonlinearity, randomness, data handling, repetition, and multimodal experiences.

Many of these software environments have pushed the boundaries of the visual arts and facilitated artists' search for an understanding of abstraction and new methods that use computers to generate art. The new generation of artists, such as Jared Tarbell, Sougwen Chung, Joshua Davis, Casey Reas, Zach Lieberman, Aaron Koblin, Karsten Schmidt, Onformative, UVA, and Fuse, among others, have produced innovative generative artworks that have attracted an increased number of spectators to the scene. Thanks to advanced audiovisual technologies, works of art that were produced with the implementation of creative codes were able to generously present all of the dynamism and spirit of the process that they contained to larger audiences.

Extending the Bounds of Creativity

Finding the one true explanation to the question of how we come up with new ideas is not an easy task. Our brains can recognize various facets, or aspects, of a single thing, circumstance, or event. The capacity to recognize diversified points of view leads to the development of original thoughts. Creativity can be defined broadly as the capacity to generate significant new ideas, forms, approaches, interpretations, etc. that deviate from conventional norms and patterns. Neuropsychologists, psychologists, and philosophers have conducted experiments to differentiate creative and uncreative people, which has helped advance our understanding of the creative process. According to Margaret Boden,[1] a research professor of cognitive science, a creative idea is one that is novel, surprising, and valuable. Attempting to delve deeper into this definition reveals the necessity of external factors, because the criteria expected to be met by creative work are not independent of an

---

[1] Boden, M. A. *The Creative Mind: Myths and Mechanisms*. Routledge (2004).

© Selçuk Artut 2023
S. Artut, *Geometric Patterns with Creative Coding*,
https://doi.org/10.1007/978-1-4842-9389-8_4

evaluation process. Likewise, Csikszentmihalyi[2] states that creativity results from the interaction between a person's thoughts and their sociocultural environment. It is a systemic phenomenon, not an individual one. In order to have a complete understanding of the social aspects of creative endeavors and achievements, it is necessary to consider not only how these things are generated internally by the mind but also how they are ingrained within a social structure.

Although creativity is most often associated with the arts, its importance in many fields has become more evident in recent years. In particular, design is one of the areas where creativity is most talked about. In addition, the importance of creativity in engineering fields is increasing day by day. Creativity allows us to see and solve problems with greater transparency and authenticity. This affords us the opportunity to gain a competitive advantage in fields where creativity is utilized. Topics such as innovation in medicine sprout ideas through creative thinking. The area of application of creativity is very vast, and its uses are unlimited.

One of the questions that is looked at is whether or not creative ability is something that can be taught. However, it is very difficult to come up with a definite opinion on this issue. Even though creativity is a highly relative concept, the outcomes of creative pursuits are still susceptible to change when viewed in the context of personal and social relationships. These outcomes are also prone to change depending on whether or not sufficient importance is given to the phenomenon of creativity. Many of us would believe that childhood is the time when people's unique creative expression is at its peak. Here, Boden makes a distinction between two creative efforts. An idea is "p-creative" if it is creative in the mind of the person in question, regardless of whether or not others have had the same idea. A thought is "h-creative" if it is "p-creative" and no one else has ever had it. According to these definitions, our childhood period of creativity would fall under the category of "p-creativeness." Nevertheless, in the long term, p-creativity will undoubtedly facilitate h-creativity. People who enjoy their individual creativity will pursue this process, and when the time comes, they will make innovative contributions to the values of the society they live in. For this reason, many open-minded educators suggest encouraging educational programs that will enable the creativity experienced naturally at a young age to endure later stages of life.

As we have seen in the preceding definitions, it is not possible to see creativity as an independent phenomenon on its own. It is only after considering the aspects of the individual, their environment, and society as a whole that it is possible to discuss the creative potential of a phenomenon. However, in addition to all of these factors, the effect that production processes have on

---

[2] Csikszentmihalyi, M. *The Systems Model of Creativity: The Collected Works of Mihaly Csikszentmihalyi.* Springer (2015).

creative output should also be taken into consideration, as it is becoming increasingly apparent with the extensive use of technological apparatus. It is possible to recognize the obvious influence that technology has had on a significant number of the works of art that we come across these days. Thus, the definition of creativity is no longer inclusive for contemporary use because it describes the activity of creativity as the embodiment of a structured thought that emerges as a result of an individual's insight and skills rather than as the activity of creativity itself, because the process of creating an artwork is no longer entirely limited to the actions of a single human being, even if that person is solely responsible for its creation. By utilizing the vast cultural and educational resources available to us, we are able to produce works of art that defy categorization and can only be attributed to the engagement of a variety of activities. In the twenty-first century, the definition of creativity as the human capacity to transform structured thoughts into physical manifestations has expanded. We can no longer consider creativity independent of technological interactions. When compared with the past, technology has, over the course of time, played a significantly more proactive role in creative actions, particularly those that take place in technological environments. Platforms such as drawing programs developed specifically for the use of artists, sound studio simulations, three-dimensional modeling, and simulation environments are now frequently utilized by artists when creating original works. At the same time, technology has extended creative reach by providing artists with tools with which to produce and conceptualize their ideas. It is no longer valid to speak of technology as a relatively passive, noninfluential force in art-making practices. Even the production of these new technologies is commonplace in today's society when the currently available technological possibilities are unable to satisfy the necessary needs. It is also possible to speak of an ever-accelerating technological development in the creation of works of art. From this point of view, it is seen that artists who design their own working environments and production methods have become widespread. The artist's comprehensive command of production technologies paves the way for the emergence of the desired rich possibilities in art production methods. While the artist is attempting to implement their artistic idea using the possibilities they have mastered, they have also begun to question the opportunities that will push the limits of these possibilities by taking the production methods one step further. This approach has enabled the artist to explore and experience production processes that expand the boundaries of creativity. Consequently, this has provided them with the chance to devise a more meaningful and unique style of artistic expression. In this regard, the most fortunate artists were those who could create their own software and hardware needs.

Musicians in the industry now have the luxury of producing their own music, but they also have access to platforms where they can design the mechanism that will build their own sound universe in any style they prefer. By embracing the possibilities of digital tools, musicians have been able to broaden their scope and develop more tailored approaches to creating music. They are now able to take their artistry to a whole new level, unrestricted by the constraints of existing production methods. Now we can talk about "digital lutherie," a new concept focused on the creation of digital musical instruments that immensely enrich musical creativity. Through digital lutherie, musicians are able to create instruments that accurately capture their desired sound, giving them greater freedom and control over the production of their music. It is revolutionizing the way we experience and create music, allowing for greater customization and personalization.

Similar to how musicians use programming paradigms to push the limits of creativity, visual artists are doing the same with the creative coding platforms. Creative coding platforms provide artists with powerful tools to explore the creative potential of code, unleashing endless possibilities for them to create exquisite works of art. Through the use of generative systems, it is now feasible to create works of art without first formulating an idea for what the work should look like. By intuitively beginning with predictions and responding to the outcomes of developed algorithms, it is possible to create works of art that reveal results that are beyond the imagination of the artist. Dialectical creativity is what I call the action that occurs when the creative process between an artist and a piece of technology goes beyond the artist's planned dreams and produces results that are intriguing.

In this process, fundamental ideas like intuition, randomness, appreciation of accidents, and experimentation play a crucial role. To fully immerse themself in the creative unity they cultivate with technology, the artist intuitively lets go of their control and awareness over the creative process. At this point, the moment of flow commences. In the same way that a pianist internalizes the notes, retrieves them from the recesses of their mind, and then performs the music in a state of flow, the artist contributes to the evolution of an artwork within the environment of creative freedom that they have achieved with the use of technology.

Flusser[3] distinguishes between structural complexity and functional complexity in an interview at the European Media Art Festival in Osnabrück, Germany, in 1988. When systems with structural complexity are examined in terms of the components they have, they have an extremely difficult and complex infrastructure, such as computers. It is possible to obtain extremely complex outcomes through the use of functionally complex systems. Flusser uses the game of chess as an illustration of functional complexity, despite the fact that the game's structural complexity is relatively straightforward. Chess is one of the most complex structures in the world in terms of its rich functional

---

[3] Vilém Flusser — 1988 interview about technical revolution. [Video]. YouTube. www.youtube.com/watch?v=1yfOcAAcoH8 (accessed December 8, 2022).

complexity for the development of games that are designed with certain strategies and presented in different ways. Nevertheless, the influence that structural and functional complexity have on creative potential is the most important aspect of Flusser's observation. Flusser demonstrates that when structural and functional complexity are combined, the creative potential of an individual or group can be greatly amplified.

The computer is an example of a technological device that possesses a very high level of structural complexity. On the other hand, the answer to the question of at what level the functional complexity of this system will occur is entirely dependent on the actions of the users. For instance, people who don't have any significant relationships outside of meeting their basic day-to-day needs while sitting in front of a computer naturally experience this complexity as extremely lacking in strength. In contrast, encounters involving dialectical creativity will have significantly greater functional complexity and potential outcomes. By effectively bringing people and technology closer together, it is possible to increase both the level of complexity and the prominence of the results achieved. As a result, the effectiveness of this system is inextricably linked to the level of participation and commitment shown by its users. The more actively people engage with the system, the greater the possibilities for change and transformation. Furthermore, the complexity of the interactions can be further elevated by allowing users to customize their own individual specifications, allowing them to create unique and personalized outcomes.

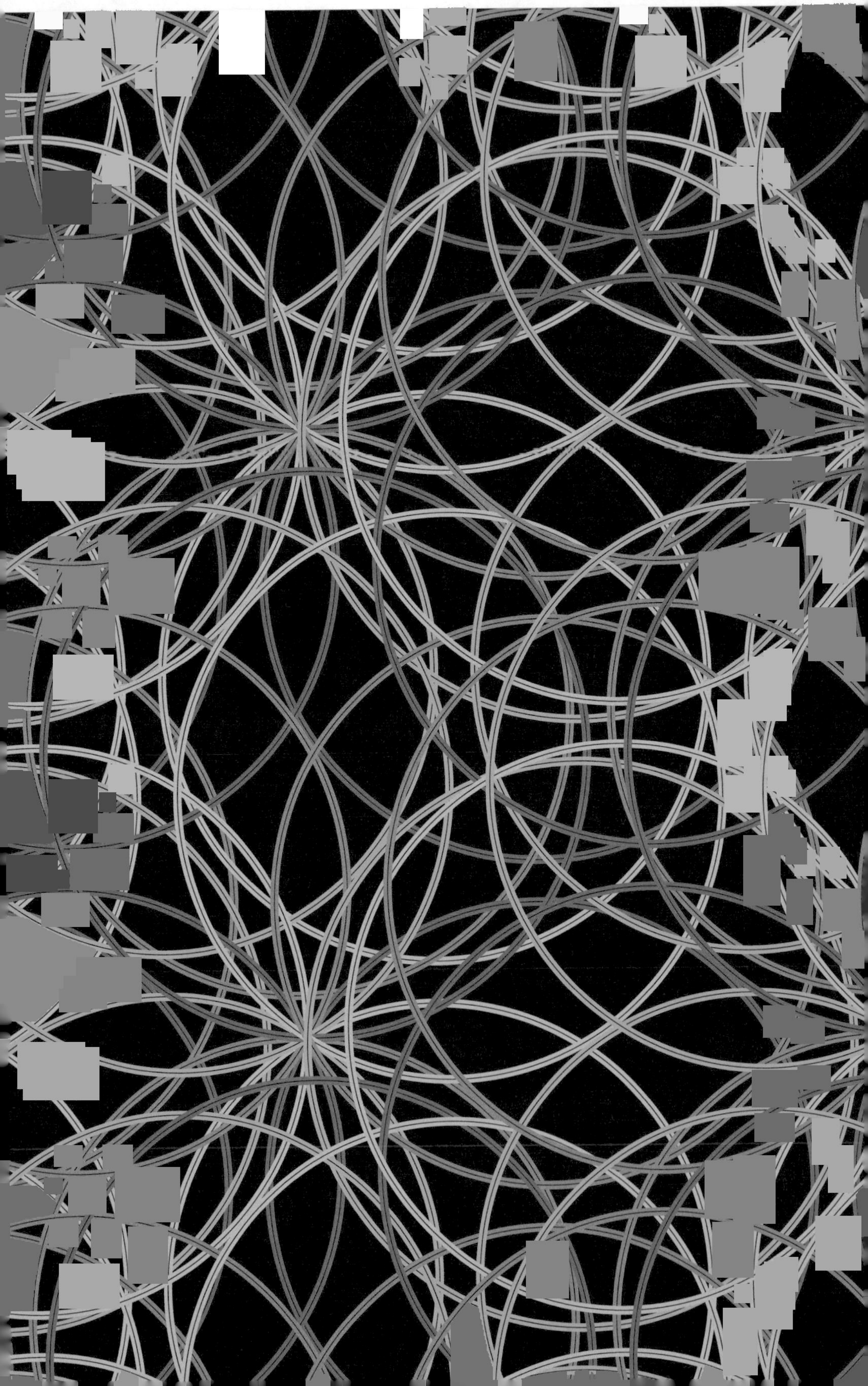

Playing with Creative Coding

What Is Creative Coding? Why Is It Different Than Conventional Programming?

Computer programming is a process of assembling a set of procedures designed to perform a certain task in a language that computers can understand. To instruct computers to carry out a given set of tasks, computer programmers might need to prefer a specific language to another. Not only is it a necessity to choose the right programming language, but programming paradigms are equally important to get a functional result. However, in comparison, the purpose of creative coding is to build something expressive rather than something useful. For this reason, creative coders may prefer to choose their programming environment according to different criteria than conventional computer programmers. Usually, creative coding prioritizes creative comfort over high efficiency. Likewise, widely used creative coding frameworks are designed to comply with a developer's creative essentials.

© Selçuk Artut 2023
S. Artut, *Geometric Patterns with Creative Coding*,
https://doi.org/10.1007/978-1-4842-9389-8_5

Creative Coding Environments

There are two main programming paradigms employed in the domain of creative coding: Visual Programming and Textual Programming.

Visual Programming (a.k.a. Node-based) enables users to build programs by manipulating program components graphically as opposed to textually. For the developers who are keen to work with interfaces and to combine different objects together to organize a complex system, Visual Programming structures are handy to start doing things very quickly.

In Textual Programming, the representation of programming processes is based on written instructions. The very first page to start writing lines of code usually starts as a tabula rasa. The developer has to foresee the needs of a project and create necessary objects from scratch if required.

The most common creative coding platforms available are

Visual Programming:

TouchDesigner

MaxMSP/Jitter

VVVV

Textual Programming:

Processing

OpenFrameworks

Cinder

p5.js

For the sake of simplicity and convenience, we will use p5.js throughout the course of the book:

p5.js is a JavaScript library for creative coding that places an emphasis on making coding accessible and inclusive for people from all walks of life, including artists, designers, educators, programmers, etc. With this in mind, this book offers an opportunity to gain a deeper understanding of geometric patterns and their possibilities through the use of p5.js (`https://p5js.org/`).

How to Use p5.js

Method 1: Using the online editor

Follow the link `https://editor.p5js.org/`.

Method 2: Using offline with an editor

Prerequisite: To run p5.js in your computer, basically you will need an editor, that is, Visual Studio Code, Sublime Text, Notepad++, or Atom.

1. Browse to the download page of p5.js (`https://p5js.org/download`) and download the p5.js complete library.

2. Unzip it to a desired location on your computer's disk space.

3. Go inside the folder that says "empty-example".

4. Edit the sketch.js file and open index.html with a browser of your preference to see the compiled results.

Optional:

If you want to see results without refreshing the browser after updating a code, you will need to install specific extensions. For instance, in Visual Code, navigate to Code (Mac)/File (Win) □ Preference □ Extensions and search for p5.vscode. After adding the extension, click the "Go Live" button in the bottom status bar to open your sketch in a browser.

Additionally, if you want to use the p5.sound library, these extensions also allow you to run a local server to work with sound.

p5.js Coding Structure

Every time a new p5.js project is initiated, two basic functions are created as default.

The **setup** function is called once as the code is compiled.

The **draw** function is repeatedly called as soon as the code is compiled. The default value is 60 fps (frame per second).

```
function setup() {
        createCanvas(400, 400);          ←          once
}

function draw() {
        background(220);                  ⟳          repeatedly
}
```

Use of Comments

Comments are essentially important in computer programming. They may serve several purposes. It is important to say that the comments are to be used by humans only. So computers will ignore those lines and compile the uncommented lines. Then why would we need them for?

Use case example:

You may need to write down a note for yourself to remember a particular use of a code.

Use case example:

When sharing a code with others, you may need to indicate some notes on particular instances.

Use case example:

You may want to ignore a block or a line of code to check for debugging.

Comment types:

Block comment: /* .... */ (using forward slash asterisk ... asterisk forward slash)

Line comment: // (using two forward slashes)

Debugging Your Sketches Using the Console

In order to check if there are any errors in your compiled sketches or try to track a variable value, you may use the console to see the "print" messages. If you are using the online editor, the console window sits under the code screen. Otherwise, you would need to look for the console section on your specific browser. Usually, it is placed under the Developer Tools section. You may use console.log or print for displaying things in the console window.

Example:

```
function setup() {
  createCanvas(400, 400);
}

function draw() {
  background(220);
  print("Hello World");
}
```

Variables

In order to create dynamic structures, we are required to have data carriers: variables that can change their value at any moment. So variables are a great tool for creating dynamic systems.

Variables are classified into two types:

Local variables: Defined within a code block and accessible only from that part of the code

Global variables: Can be accessed from anywhere in the code

Usually in most common programming languages such as C++ and Java, in order to reduce the computer memory load, you would need to declare a variable type. But in p5.js, all you need to do is to declare **let** as a variable type, and the compiler will take care of the rest for you.

Variable examples:

```
let a = 5;               // a is a number
let t = "hello world";   // a is a text
let b = true;            // a is a boolean
let c = color(255, 204, 0);  // a is an RGB color
```

Loops

Repetition is one of the things computers do best without too much effort. In order to create an operation to run over and over, we use the structure called loops in programming. There are several types of loops for various needs. The most common one that we will introduce here is the for loop. See the following example:

```
for (let i = 0; i < 3; i++) {
  console.log(i);
}
```

In a for loop, we start with declaring a variable to iterate during the repetition. It is called the initialization. In the preceding code, for its initial setting, the i value is set to be a number that is equal to zero.

Second, we specify a condition for the loop to satisfy until it ends repeating. Third, each time the loop repeats, the i value will be either increased or decreased by a specific value. This situation determines the repeat times since the change will determine the number of times to reach a condition to break the loop.

Iteration 1:

i=0. Since 0 is less than 3, we grant the loop to proceed with its content.

```
for (let i = 0; i < 3; i++) {
  console.log(i);
}
```

This will print zero on the console, add one due to the i++ operation, and call for the next loop.

Iteration 2:

i=1. Since 1 is less than 3, we grant the loop to proceed with its content.

```
for (let i = 0; i < 3; i++) {
  console.log(i);
}
```

This will print one on the console, add one due to the i++ operation, and call for the next loop.

Iteration 3:

i=2. Since 2 is less than 3, we grant the loop to proceed with its content.

```
for (let i = 0; i < 3; i++) {
  console.log(i);
}
```

This will print two on the console, add one due to the i++ operation, and call for the next loop.

Iteration 4:

i=3. Since 3 is not less than 3, the loop breaks.

As we observed here, we had three repeating iterations. Loops are one of the most essential blocks of code structures that we use in generating patterns with ease.

Object-Oriented Programming: Building a Class

Object-oriented programming is based on the principle that reusable objects can be employed to create a general structure. In this book, classes will serve as blueprints for generating motifs as tiles. These building blocks will then allow us to render tessellations by tiling multiple motifs together.

Example:

```
// Motif class
class Motif {
  constructor() {
  //this function is called once when an instance is initiated
  }

  draw() {
  //this is a user defined function
  }
}
let motif;

function setup() {
  createCanvas(800, 800);
  //initiate an instance
  motif = new Motif();
}
```

```
function draw() {
  background(255);
  motif.draw();
}
```

Cartesian Coordinates in a Computer Graphics Environment

The usual convention is to set the Cartesian coordinate system placed (0,0) in the center with the y-axis pointing up and the x-axis pointing to the right (in the positive direction, negative down and to the left). The coordinate system for pixels in a computer window, however, is reversed along the y-axis. (0,0) can be found at the top left with the positive direction to the right horizontally and down vertically. That means that the first quadrant where x and y are positive is in the bottom right.

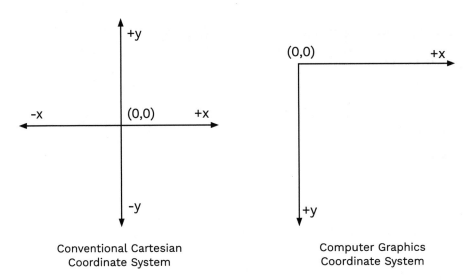

Conventional Cartesian
Coordinate System

Computer Graphics
Coordinate System

Drawing Basic Shapes

p5.js enables the developers to use some built-in basic shapes to work with such as point, circle, ellipse, rectangle, square, line, arc, and triangle.

Here are some examples and uses of such functions.

Point:

Draws a single pixel at a given coordinate location.

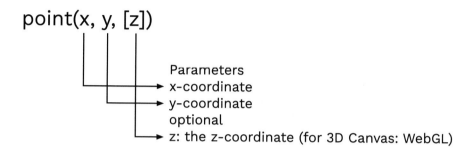

Example:

```
//canvas 200 px width, 200 px height
point(50, 30);
```

Circle:

Draws a circle with a given diameter centered at a given coordinate location.

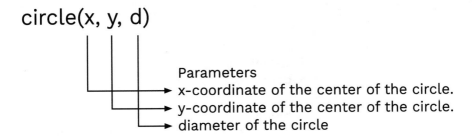

circle(x, y, d)

Parameters
x-coordinate of the center of the circle.
y-coordinate of the center of the circle.
diameter of the circle

Example:

```
//canvas 200 px width, 200 px height
circle(100,50,30);
```

Ellipse:

Draws an ellipse with a given width and height centered at a given coordinate location.

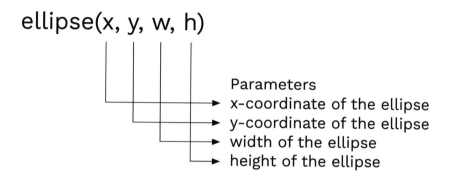

Parameters
x-coordinate of the ellipse
y-coordinate of the ellipse
width of the ellipse
height of the ellipse

Example:

```
//canvas 200 px width, 200 px height
ellipse(50, 50, 30, 20);
```

Rectangle:

Draws a rectangle with a given width and height with its upper-left corner placed at a given coordinate location.

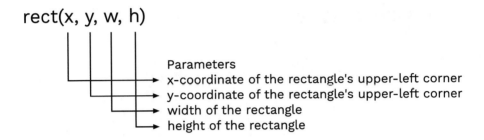

Parameters
x-coordinate of the rectangle's upper-left corner
y-coordinate of the rectangle's upper-left corner
width of the rectangle
height of the rectangle

Example:

```
//canvas 200 px width, 200 px height
rect(50, 50, 40, 60);
```

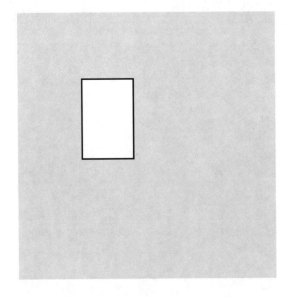

Square:

Draws a square with a given width and height with its upper-left corner placed at a given coordinate location.

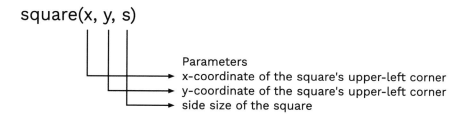

square(x, y, s)

Parameters
x-coordinate of the square's upper-left corner
y-coordinate of the square's upper-left corner
side size of the square

Example:

```
//canvas 200 px width, 200 px height
square(50, 50, 40);
```

Line:

Draws a line from a beginning point to an ending point.

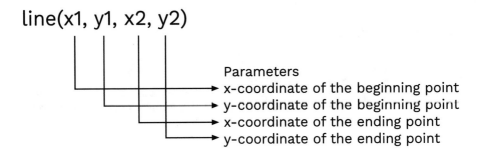

line(x1, y1, x2, y2)

Parameters
x-coordinate of the beginning point
y-coordinate of the beginning point
x-coordinate of the ending point
y-coordinate of the ending point

Example:

```
//canvas 200 px width, 200 px height
line(20,20,160,160);
```

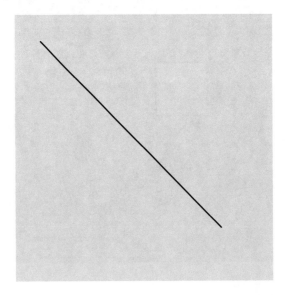

Arc:

Draws an arc with a given width, height, and interval of a degree.

arc(x, y, w, h, start, stop)

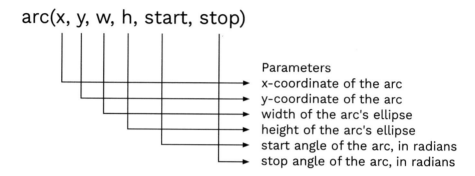

Parameters
x-coordinate of the arc
y-coordinate of the arc
width of the arc's ellipse
height of the arc's ellipse
start angle of the arc, in radians
stop angle of the arc, in radians

Example:

```
//canvas 200 px width, 200 px height
arc(100,100,60,80,0, HALF_PI);
```

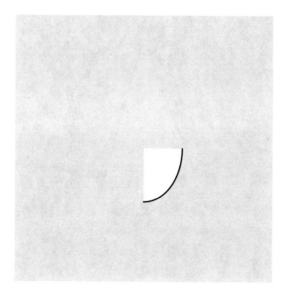

Triangle:

Draws a triangle in between three points.

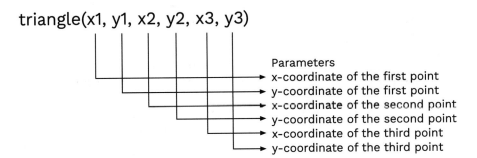

Parameters
x-coordinate of the first point
y-coordinate of the first point
x-coordinate of the second point
y-coordinate of the second point
x-coordinate of the third point
y-coordinate of the third point

Example:

```
//canvas 200 px width, 200 px height
triangle(100,20,140,160, 60,120);
```

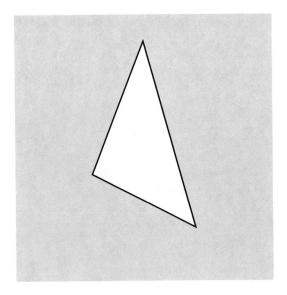

Drawing Complex Shapes

```
function setup() {
  createCanvas(200, 200);
}

function draw() {
  background(220);
  beginShape();
  vertex(100, 20);
  vertex(30, 45);
  vertex(80, 40);
  vertex(50, 160);
  vertex(150, 160);
  vertex(80, 110);
  endShape(CLOSE);
}
```

Transformations

Transformations in computer graphics are very important in terms of the flexibility they provide. Basically, transformation refers to the mathematical operations that are applied on a graphical shape to manipulate its location, size, or orientation on a screen. Some most commonly used types of transformations are

Translation

Rotation

Scaling

Every canvas we see on the screen is structured with a computer Cartesian system with an invisible set of coordinates. If you'd like to create a circle at a specific position, you would need to specify its location and its diameter:

```
circle(40,50,20); //circle(x-pos, y-pos, diameter)
```

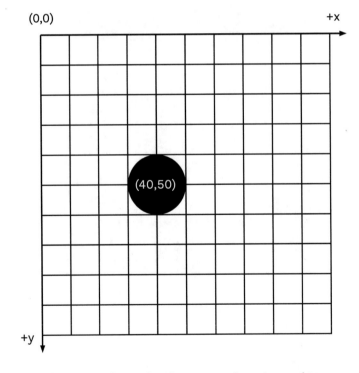

With the aid of the transformation functions, there is another way of placing your shape on a canvas. You may use the translation function to change its location as such:

```
translate(40,50);
circle(0,0,20); //circle(x-pos, y-pos, diameter)
```

This will give you the same result, but the process that is running underneath is slightly different. In transformation, you may think of it as you are moving the canvas rather than the shape itself. It is very much like shifting a paper's location and rotation while you are painting physically to give it a comfort for yourself.

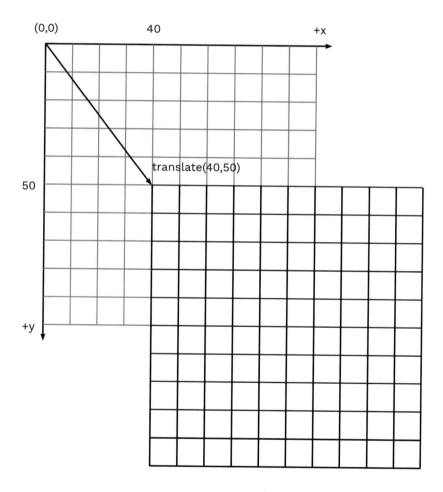

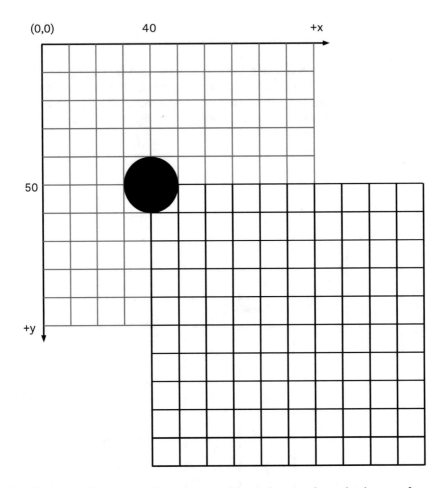

But there may be times when you would need to apply multiple transformation functions on multiple shapes. In that case, we would need to introduce push() and pop() functions in here. Push stores the current status of the coordinate system to a specific setting, and pop returns the system back to its original status.

Let's imagine that we would like to have multiple circles on canvas at different locations. One of the main advantages of applying a transformation function is that it applies to a shape regardless of its internal settings:

```
function setup() {
  createCanvas(400, 400);
  noLoop();
}

function draw() {
  background(220);
  for(let i = 0; i < 150; i++){
        push();
        translate(random(width),random(height));
        circle(0,0,random(10,40));
        pop();
  }
}
```

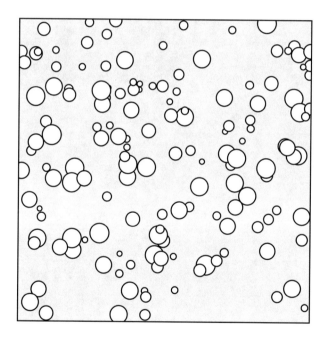

Generating a Basic Geometric Pattern

Use of Hand Tools vs. Creative Coding Methods

Generating a Basic Geometric Pattern with Hand Tools

Method 1: Hand-drawing using a ruler and compass

S. Artut, *Geometric Patterns with Creative Coding*,
https://doi.org/10.1007/978-1-4842-9389-8_6

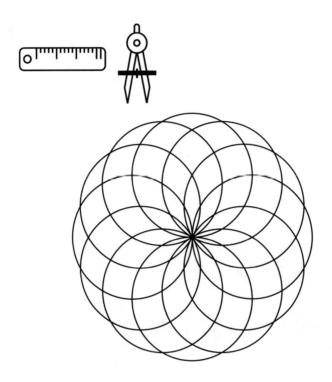

Division of a Circle in 6 and 12 Equal Sections

Step 1: Draw a straight line with a pencil.

Step 2: Use a compass to make a circle, and place its center on the line.

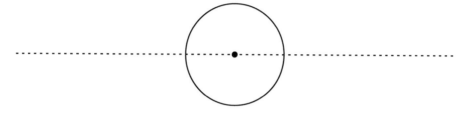

Step 3: Draw two other circles to the left and the right, and center their origin adjacent to the previous circle.

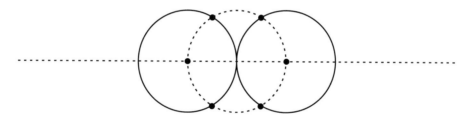

Now we have an equal-sided hexagon.

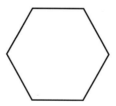

Step 4: Draw six circles in total at the intersection points.

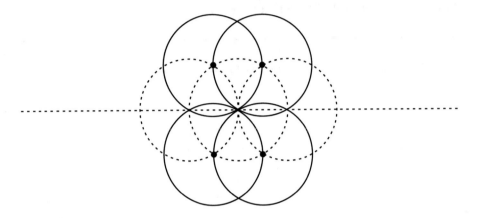

Step 5: Draw lines across the intersection points.

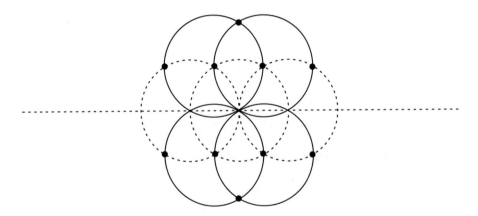

Step 6: Draw circles around twelve intersection points on the original circle.

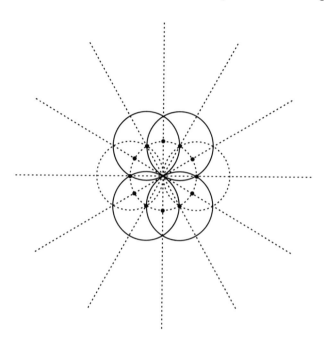

Now we have twelve equal sections on a circle.

Step 7: Erase all construction lines, and strengthen your strokes.

Generating a Basic Geometric Pattern with Creative Coding

Method 2: Using creative coding skills

Generating a Basic Geometric Pattern Using Creative Coding

Step 1: Let's start with creating a basic sketch.

Canvas size: width, 400 px; height, 400 px

The background is white.

```
function setup() {
    createCanvas(400, 400);
}
function draw() {
    background(255);
}
```

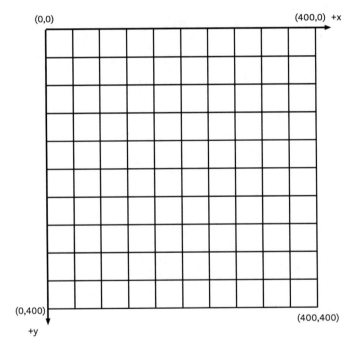

Step 2: Now we will draw a circle with no fill on our canvas.

```
function setup() {
    createCanvas(400, 400);
    noFill();
}
function draw() {
    background(255);
    circle(0,0,120);
}
```

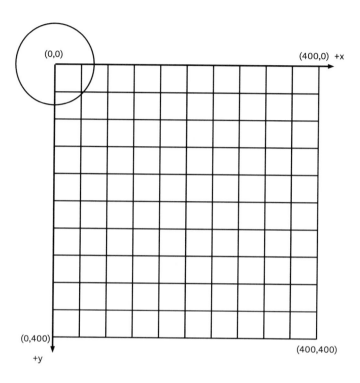

Step 3: Let's center our circle on the canvas using the translate function.

We may gather the width and height values of a canvas dynamically by calling the built-in width and height functions.

```
function setup() {
    createCanvas(400, 400);
    noFill();
}
function draw() {
    background(255);
    push();
        translate(width*0.5,height*0.5);
        circle(0,0,120);
    pop();
}
```

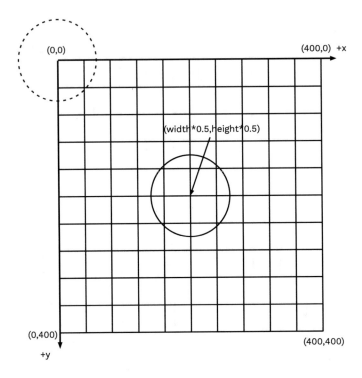

Step 4: Let's shift our circle to the right with a distance of its radius. Remember that 120 px is the diameter.

We will later have 12 copies in total of this circle and 30 multiple degrees rotated around the canvas's center.

```
function setup() {
    createCanvas(400, 400);
    noFill();
}
function draw() {
    background(255);
    push();
        translate(width*0.5,height*0.5);
        push();
            translate(60,0);
            circle(0,0,120);
        pop();
    pop();
}
```

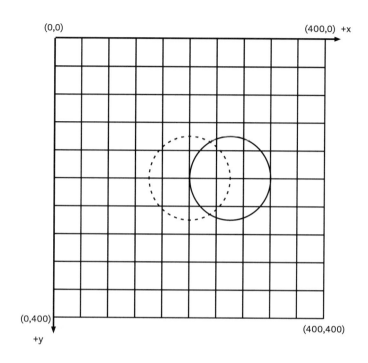

Step 5: Now we will use a for loop to get 12 copies of the circle. Be aware of the angle mode. By default, it is set to RADIANS, so we apply angleMode(DEGREES).

```
function setup() {
    createCanvas(400, 400);
    angleMode(DEGREES);
    noFill();
}
function draw() {
    background(255);
    push();
        translate(width*0.5,height*0.5);
        for(let i = 0; i < 12; i++){
        push();
            rotate(i*30);
            translate(60,0);
            circle(0,0,120);
        pop();
        }
    pop();
}
```

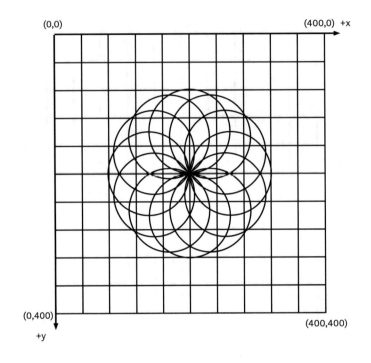

Transformation order is important!!!

In order to simplify, here below you will find three iterations only First Rotate then Translate.

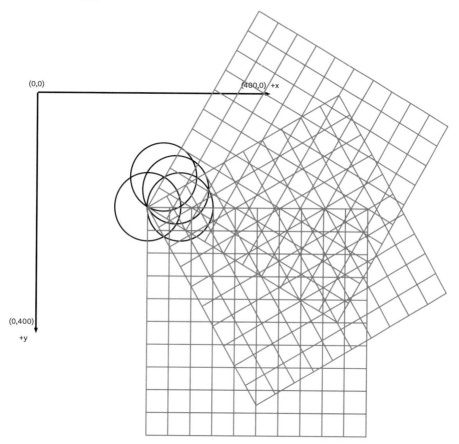

Transformation order is important!!!

In order to simplify, here below you will find three iterations only First Translate then Rotate.

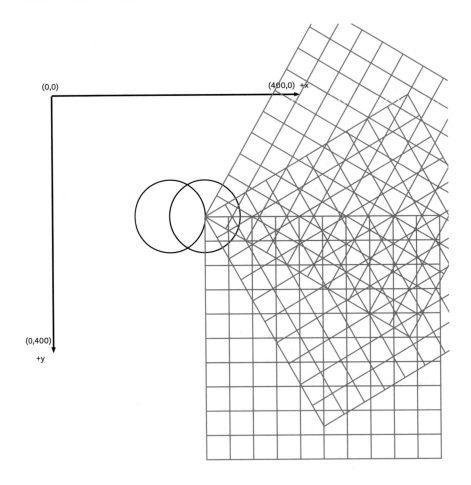

Trigonometry Basics

The three fundamental functions in trigonometry are sine, cosine, and tangent.

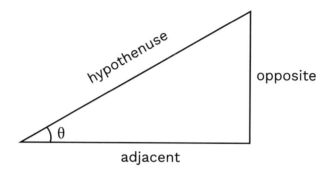

$\sin(\theta)$ = opposite/hypotenuse

$\cos(\theta)$ = adjacent/hypotenuse

$\tan(\theta)$ = opposite/adjacent

Units of Measuring Angles

Degrees are a unit of measurement, each counted as 1 of 360 equal slices of a circle. A full circle would be a total of 360 degrees.

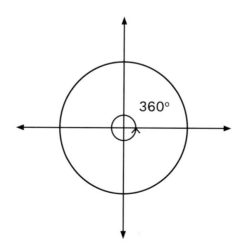

Radian is related to the diameter of a circle. The full circle's angle is 2π radians.

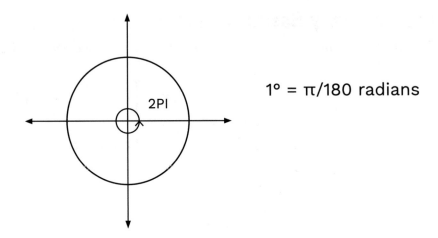

$$1° = \pi/180 \text{ radians}$$

To use "degrees" instead of the default "radians" mode, you need to apply the angleMode(DEGREES) function.

Law of Sines

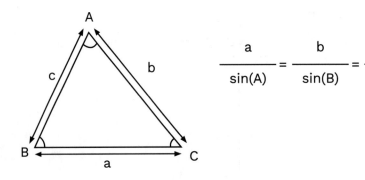

$$\frac{a}{\sin(A)} = \frac{b}{\sin(B)} = \frac{c}{\sin(C)}$$

Workflows on Generating Geometric Patterns with Creative Coding

© Selçuk Artut 2023
S. Artut, *Geometric Patterns with Creative Coding*,
https://doi.org/10.1007/978-1-4842-9389-8_7

Generating a Geometric Pattern Workflow #1

Observe the geometric pattern, and analyze it to distinguish its constituent repeating motif.

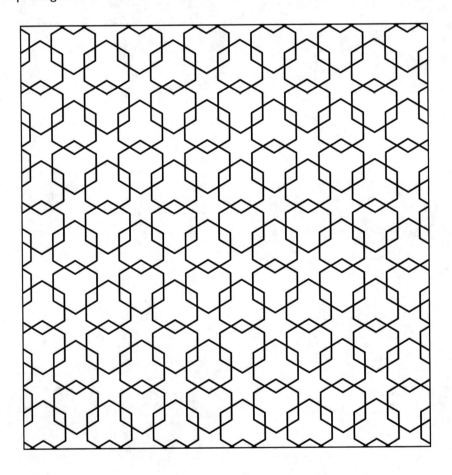

The Motif

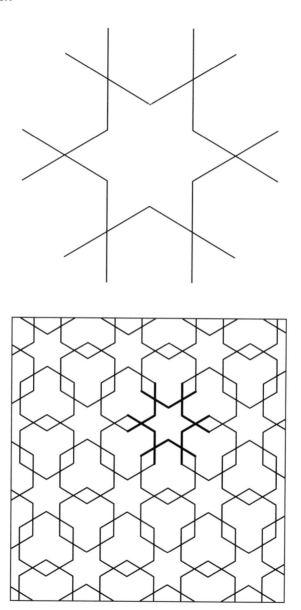

Analyzing the Constructive Elements

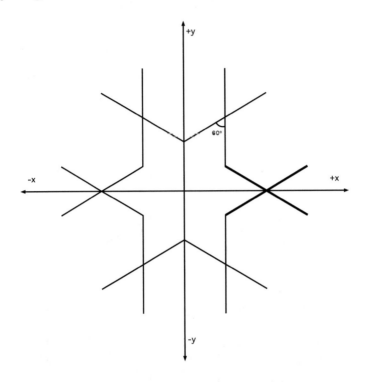

Calculating Angles and Vertex Points

Step 1: Let's find the vertex points of the constructive element.

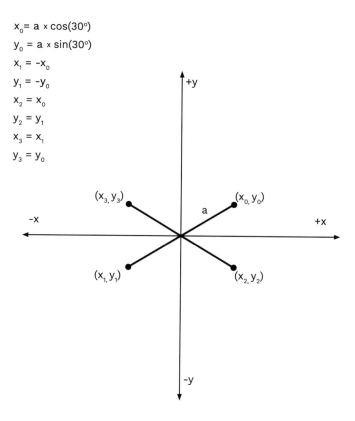

$x_0 = a \times \cos(30°)$

$y_0 = a \times \sin(30°)$

$x_1 = -x_0$

$y_1 = -y_0$

$x_2 = x_0$

$y_2 = y_1$

$x_3 = x_1$

$y_3 = y_0$

Generating the Motif

Step 2: Let's start with drawing the constructive element. We have two intersecting lines as shapes.

```
//scale factor
let a = 60;

function setup() {
      createCanvas(400, 400);
      angleMode(DEGREES);
      noFill();
      noLoop();
}

function draw() {
      let x0,y0,x1,y1,x2,y2,x3,y3;
      push();
            translate(width*0.5, height*0.5);
            //move to the right by its width size
            //line one
            beginShape();
            x0 = a * cos(30);
            y0 = a * sin(30);
            vertex(x0,y0);
            x1 = -1 * x0;
            y1 = -1 * y0;
            vertex(x1,y1);
            endShape();
            //line two
            beginShape();
            x2 = x0;
            y2 = y1;
            vertex(x2,y2);
            x3 = x1;
            y3 = y0;
            vertex(x3,y3);
            endShape();
      pop();
}
```

Step 3: Using the constructive element, we may generate the motif by apply-ing transformation functions.

Algorithm: Translate the constructive element to right with the distance equal to its width and then rotate around the center six times in a loop.

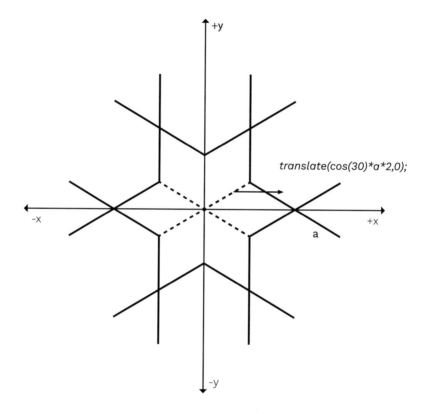

Step 4: We will use a for loop with six iterations to generate the motif. Remember, the order of transformation is important!

```
//scale factor
let a = 60;

function setup() {
    createCanvas(400, 400);
    angleMode(DEGREES);
    noFill();
    noLoop();
}

function draw() {
    let x0,y0,x1,y1,x2,y2,x3,y3;
    push();
        translate(width*0.5, height*0.5);
        for(let i=0; i<6; i++){
            push();
                rotate(i*60);
                //move to the right by its width size
                translate(cos(30)*a*2,0);
                //line one
                beginShape();
                x0 = a * cos(30);
                y0 = a * sin(30);
                vertex(x0,y0);
                x1 = -1 * x0;
                y1 = -1 * y0;
                vertex(x1,y1);
                endShape();
                //line two
                beginShape();
                x2 = x0;
                y2 = y1;
                vertex(x2,y2);
                x3 = x1;
                y3 = y0;
                vertex(x3,y3);
                endShape();
            pop();
        }
    pop();
}
```

Analyzing the Tessellation

Step 5: We need to calculate the dx, dy, and doff values in the placement.

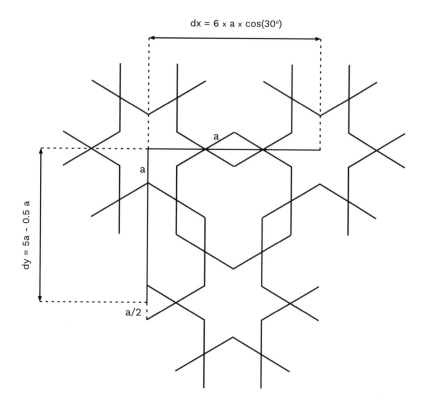

Tessellation Code

```
// Motif class
class Motif {
  constructor(a) {
      this.a = a;
  }

  display() {
      let x0, y0, x1, y1, x2, y2, x3, y3;
      for (let i = 0; i < 6; i++) {
          push();
              rotate(i * 60);
              translate(cos(30) * this.a * 2, 0);
              //line one
              beginShape();
              x0 = this.a * cos(30);
              y0 = this.a * sin(30);
              vertex(x0, y0);
              x1 = -1 * x0;
              y1 = -1 * y0;
              vertex(x1, y1);
              endShape();
              //line two
              beginShape();
              x2 = x0;
              y2 = y1;
              vertex(x2, y2);
              x3 = x1;
              y3 = y0;
              vertex(x3, y3);
              endShape();
          pop();
      }
  }
}
```

```
//scale factor
let a = 24;
let motif = new Motif(a);
let nRow;
let nCol;
let dx, dy;

function setup() {
  createCanvas(800, 800);
  angleMode(DEGREES);
  noFill();
  noLoop();

  dx = 6 * a * cos(30);
  dy = 4.5 * a;
  doff = 0.5 * dx;

  //approximate the nRow and nCol values
  nCol = ceil(width / dx);
  nRow = ceil(height / dy);
}

function draw() {
  for (let c = 0; c < nCol; c++) {
    for (let r = 0; r < nRow; r++) {
      push();
        if (r % 2 == 0) {
          //columns 0,2,4
          translate(doff, 0);
        }
        translate(c*dx,r*dy);
        motif.display();
      pop();
    }
  }
}
```

Generating a Geometric Pattern Workflow #2

Observe the geometric pattern, and analyze it to distinguish its constituent repeating motif.

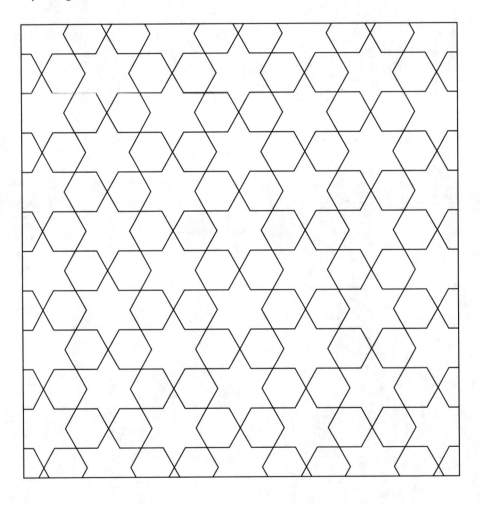

The Motif

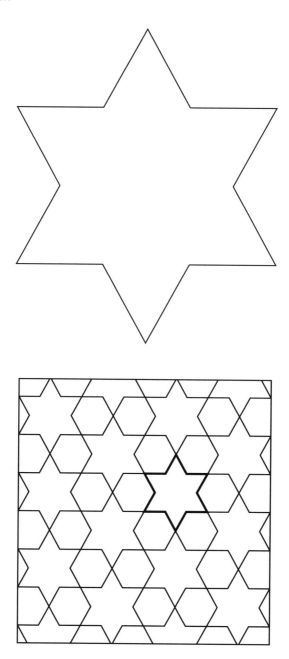

Analyzing the Constructive Elements

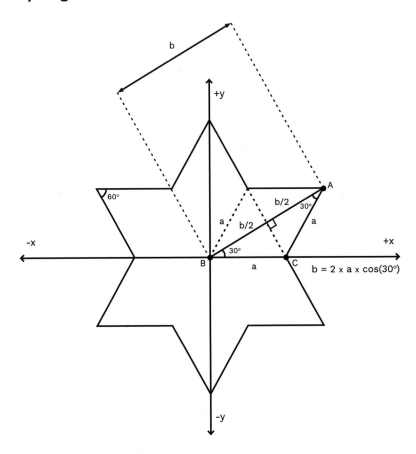

Calculating Angles and Vertex Points

Step 1: Let's find the vertex points of the constructive element.

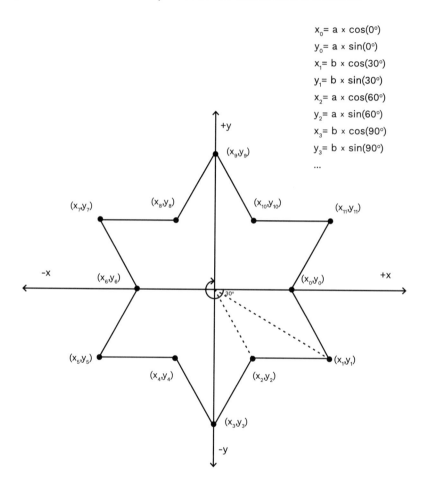

$$x_0 = a \times \cos(0°)$$
$$y_0 = a \times \sin(0°)$$
$$x_1 = b \times \cos(30°)$$
$$y_1 = b \times \sin(30°)$$
$$x_2 = a \times \cos(60°)$$
$$y_2 = a \times \sin(60°)$$
$$x_3 = b \times \cos(90°)$$
$$y_3 = b \times \sin(90°)$$
...

Generating the Motif

```
//scale factor
let a = 80;
let b;

function setup() {
      createCanvas(400, 400);
      angleMode(DEGREES);
      b = 2 * a * cos(30);
      noFill();
      noLoop();
}
function draw() {
      let x,y;
      push();
          translate(width*0.5, height*0.5);
          beginShape();
          for(let i=0;i<12;i++){
              if(i%2==0){
                      x = a*cos(30*i);
                      y = a*sin(30*i);
              }else{
                      x = b*cos(30*i);
                      y = b*sin(30*i);
              }
              vertex(x,y);
          }
          endShape(CLOSE);
      pop();
}
```

Analyzing the Tessellation

Step 2: We need to calculate the dx, dy, and dyoff values in the placement.

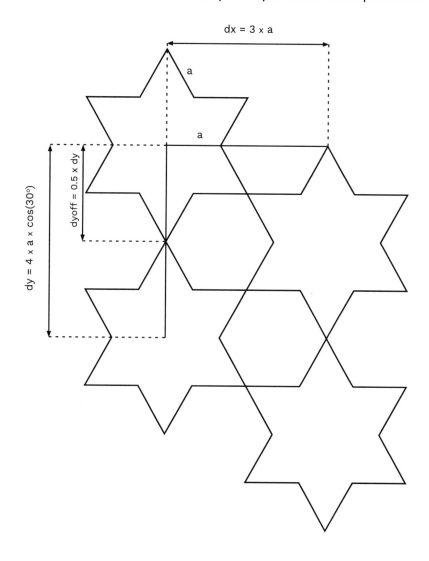

Tessellation Code

```
// Motif class
class Motif {
  constructor(a) {
    this.a = a;
  }

  display() {
    let b = 2 * this.a * cos(30);
    beginShape();
    for(let i=0;i<12;i++){
        let x,y;
        if(i%2==0){
          x = this.a*cos(30*i);
          y = this.a*sin(30*i);
        }else{
          x = b*cos(30*i);
          y = b*sin(30*i);
        }
        vertex(x,y);
      }
      endShape(CLOSE);
  }
}

//scale factor
let a = 40;
let motif = new Motif(a);
let nRow;
let nCol;
let dx, dy;

function setup() {
  createCanvas(800, 800);
  angleMode(DEGREES);
  noFill();
  noLoop();

  dx = 3 * a;
  dy = 4 * a * cos(30);

  //approximate the nRow and nCol values
  nRow = ceil(height / dy);
  nCol = ceil(width / dx);

}
```

```
function draw() {
  for (let c = 0; c < nCol; c++) {
    for (let r = 0; r < nRow; r++) {
      push();
        translate(dx * c, dy * r);
        if(c%2==0){
          //columns 0,2,4
          translate(0,dy*0.5);
        }
        motif.display();
      pop();
    }
  }
}
```

Generating a Geometric Pattern Workflow #3

Observe the geometric pattern, and analyze it to distinguish its constituent repeating motif.

The Motif

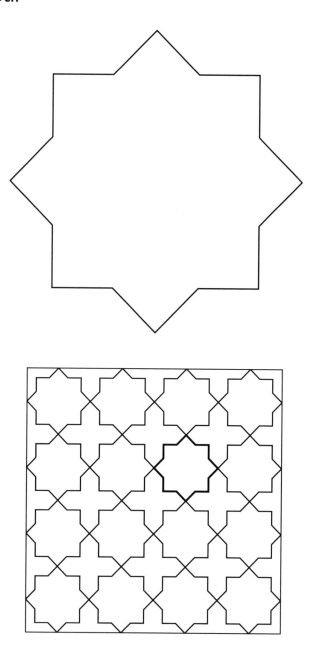

Analyzing the Constructive Elements

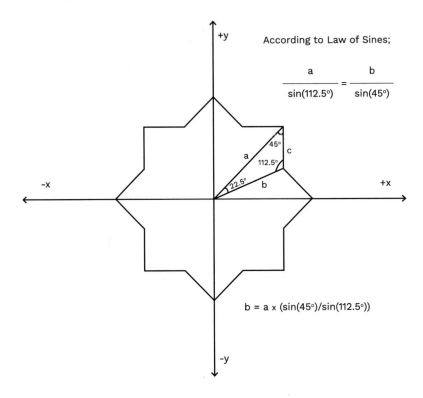

According to Law of Sines;

$$\frac{a}{\sin(112.5°)} = \frac{b}{\sin(45°)}$$

$b = a \times (\sin(45°)/\sin(112.5°))$

Calculating Angles and Vertex Points

Step 1: Let's find the vertex points of the constructive element.

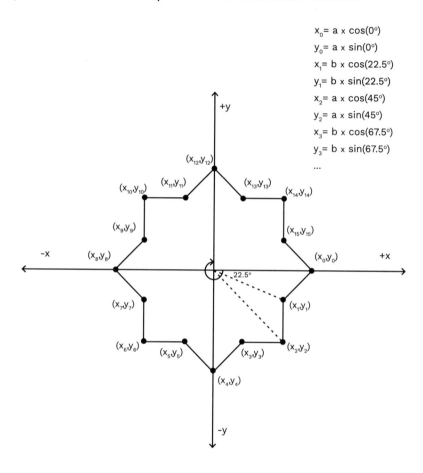

$x_0 = a \times \cos(0°)$
$y_0 = a \times \sin(0°)$
$x_1 = b \times \cos(22.5°)$
$y_1 = b \times \sin(22.5°)$
$x_2 = a \times \cos(45°)$
$y_2 = a \times \sin(45°)$
$x_3 = b \times \cos(67.5°)$
$y_3 = b \times \sin(67.5°)$
...

Generating the Motif

```
//scale factor
let a = 120;
let b;

function setup() {
  createCanvas(400, 400);
  angleMode(DEGREES);
  noFill();
  noLoop();
  b = a * (sin(45) / sin(112.5));
}
function draw() {
  let x,y;
  push();
    translate(width*0.5, height*0.5);
    rotate(22.5);
    beginShape();
    for (let i = 0; i < 8; i++) {
      let sx = cos(i*45) * b;
      let sy = sin(i*45) * b;
      vertex(sx, sy);
      sx = a * cos(i *45 + 45*0.5);
      sy = a * sin(i *45 + 45*0.5);
      vertex(sx, sy);
    }
    endShape(CLOSE);
  pop();
}
```

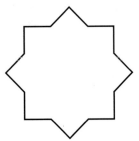

Analyzing the Tessellation

Step 2: We need to calculate the dx and dy values in the placement.

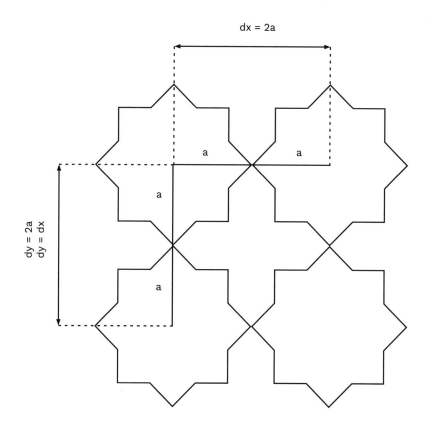

Tessellation Code

```
// Motif class
class Motif {
  constructor(a) {
    this.a = a;
  }

  display() {
    push();
      rotate(22.5);
      let b = this.a * (sin(45)/sin(112.5));
      beginShape();
      for (let i = 0; i < 8; i++) {
        let sx = cos(i*45) * b;
        let sy = sin(i*45) * b;
        vertex(sx, sy);
        sx = this.a * cos(i *45 + 45*0.5);
        sy = this.a * sin(i *45 + 45*0.5);
        vertex(sx, sy);
      }
      endShape(CLOSE);
    pop();
  }
}

//scale factor
let a = 60;
let motif = new Motif(a);
let nRow;
let nCol;
let dx, dy;

function setup() {
  createCanvas(800,800);
  angleMode(DEGREES);
  noFill();
  noLoop();

  let b = a * (sin(45)/sin(112.5));

  dx = 2*a;
  dy = dx;

  //approximate the nRow and nCol values
  nRow = ceil(height / dy);
  nCol = ceil(width / dx);

}
```

```
function draw() {
  for (let c = 0; c < nCol; c++) {
    for (let r = 0; r < nRow; r++) {
      push();
        translate(c * dx, r * dy);
        motif.display();
      pop();
    }
  }
}
```

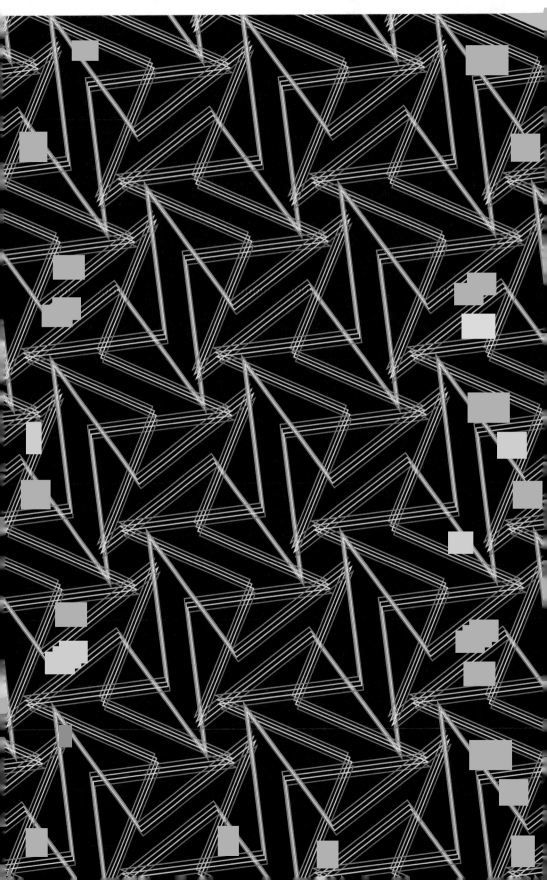

Generating a Geometric Pattern Workflow #4

Observe the geometric pattern, and analyze it to distinguish its constituent repeating motif.

The Motif

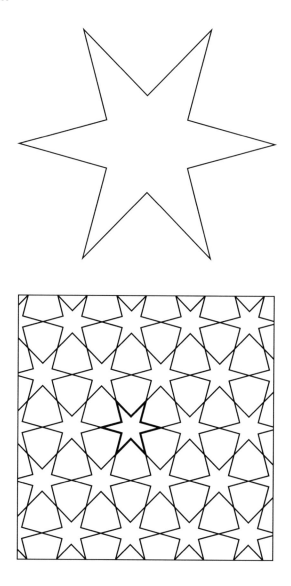

Analyzing the Constructive Elements

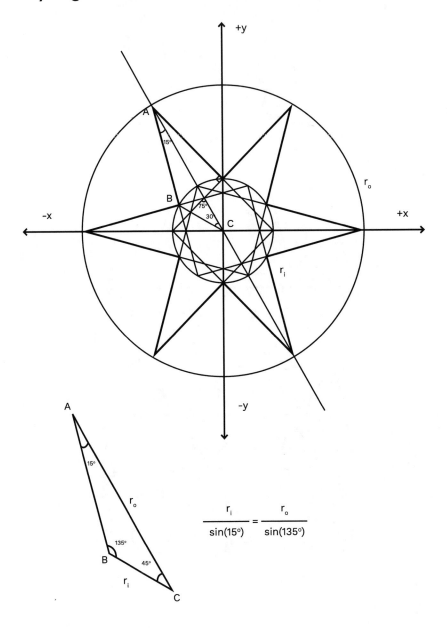

Calculating Angles and Vertex Points

Step 1: Let's find the vertex points of the constructive element.

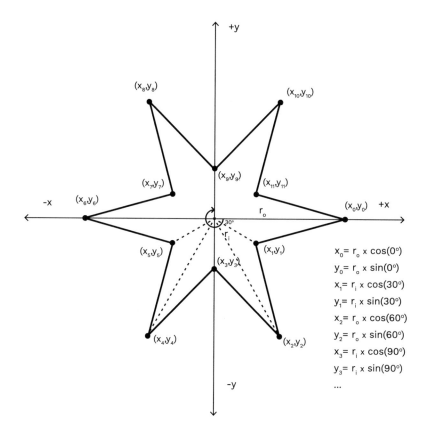

$x_0 = r_o \times \cos(0°)$
$y_0 = r_o \times \sin(0°)$
$x_1 = r_i \times \cos(30°)$
$y_1 = r_i \times \sin(30°)$
$x_2 = r_o \times \cos(60°)$
$y_2 = r_o \times \sin(60°)$
$x_3 = r_i \times \cos(90°)$
$y_3 = r_i \times \sin(90°)$
...

Generating the Motif

```
let a = 48; //inner Radius, scale factor
let b; //outer Radius

function setup() {
  createCanvas(400, 400);
  angleMode(DEGREES);
  noFill();
  noLoop();
  b = a * (sin(135) / sin(15));
}

function draw() {
  let angle = 30;

  push();
      translate(width*0.5,height*0.5);
      beginShape();
      for (let i = 0; i < 12; i++) {
          let sx,sy;
          if(i%2==0){
              sx = cos(i*angle) * b;
              sy = sin(i*angle) * b;
          }else{
              sx = cos(i*angle) * a;
              sy = sin(i*angle) * a;
          }
          vertex(sx, sy);
      }
      endShape(CLOSE);
  pop();
}
```

Analyzing the Tessellation

Step 2: We need to calculate the dx, dy, and dxoff values in the placement.

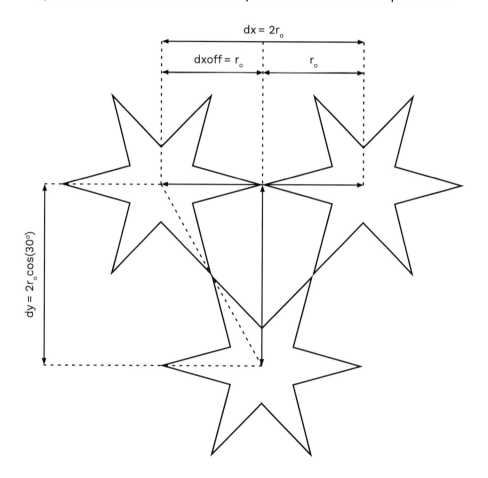

Tessellation Code

```
// Motif class
class Motif {
  constructor(r) {
    this.a = r; //inner Radius
    this.b = r * (sin(135) / sin(15)); //outer Radius
  }

  display() {
      let angle = 30;
      beginShape();
      for (let i = 0; i < 12; i++) {
          let sx,sy;
          if (i%2==0){
              sx = cos(i*angle) * this.b;
              sy = sin(i*angle) * this.b;
          } else {
              sx = cos(i*angle) * this.a;
              sy = sin(i*angle) * this.a;
          }
          vertex(sx, sy);
      }
      endShape(CLOSE);
  }
}

let a = 16; //inner Radius, scale factor
let b; //outer Radius
let dx, dy;
let nRow;
let nCol;

function setup() {
  createCanvas(800, 800);
  angleMode(DEGREES);
  noFill();
  noLoop();
  b = a * (sin(135) / sin(15));
  dx = 2 * b;
  dy = 2 * b * cos(30);
  //approximate the nRow and nCol values
  nRow = ceil(height / dy);
  nCol = ceil(width / dx);
}
```

```
function draw() {
    let motif = new Motif(a);
    for (let r = 0; r < nRow; r++) {
        for (let c = 0; c < nCol; c++) {
            push();
                if (r % 2 == 0) {
                    //rows 0,2,4,6
                    translate(c * dx,   r * dy);
                } else {
                    //rows 1,3,5,7
                    translate(c * dx  + b,   r * dy);
                }
                motif.display();
            pop();
        }
    }
}
```

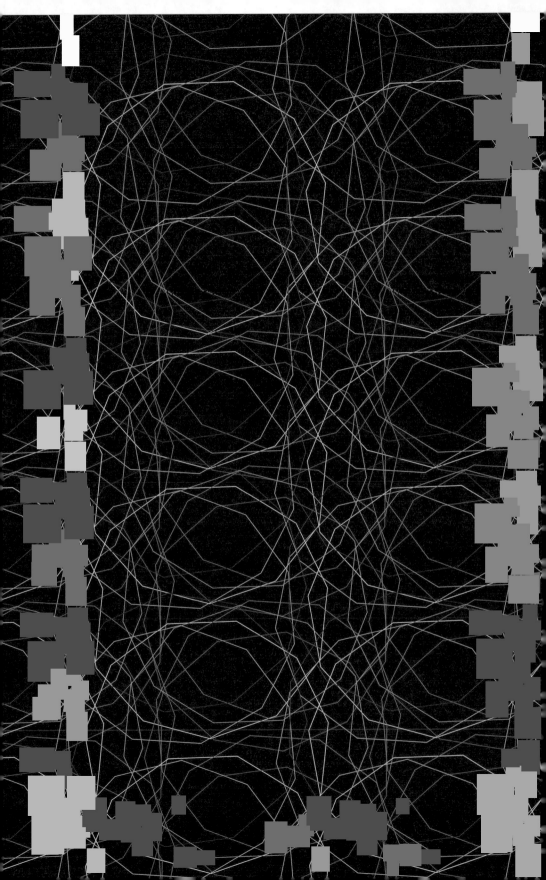

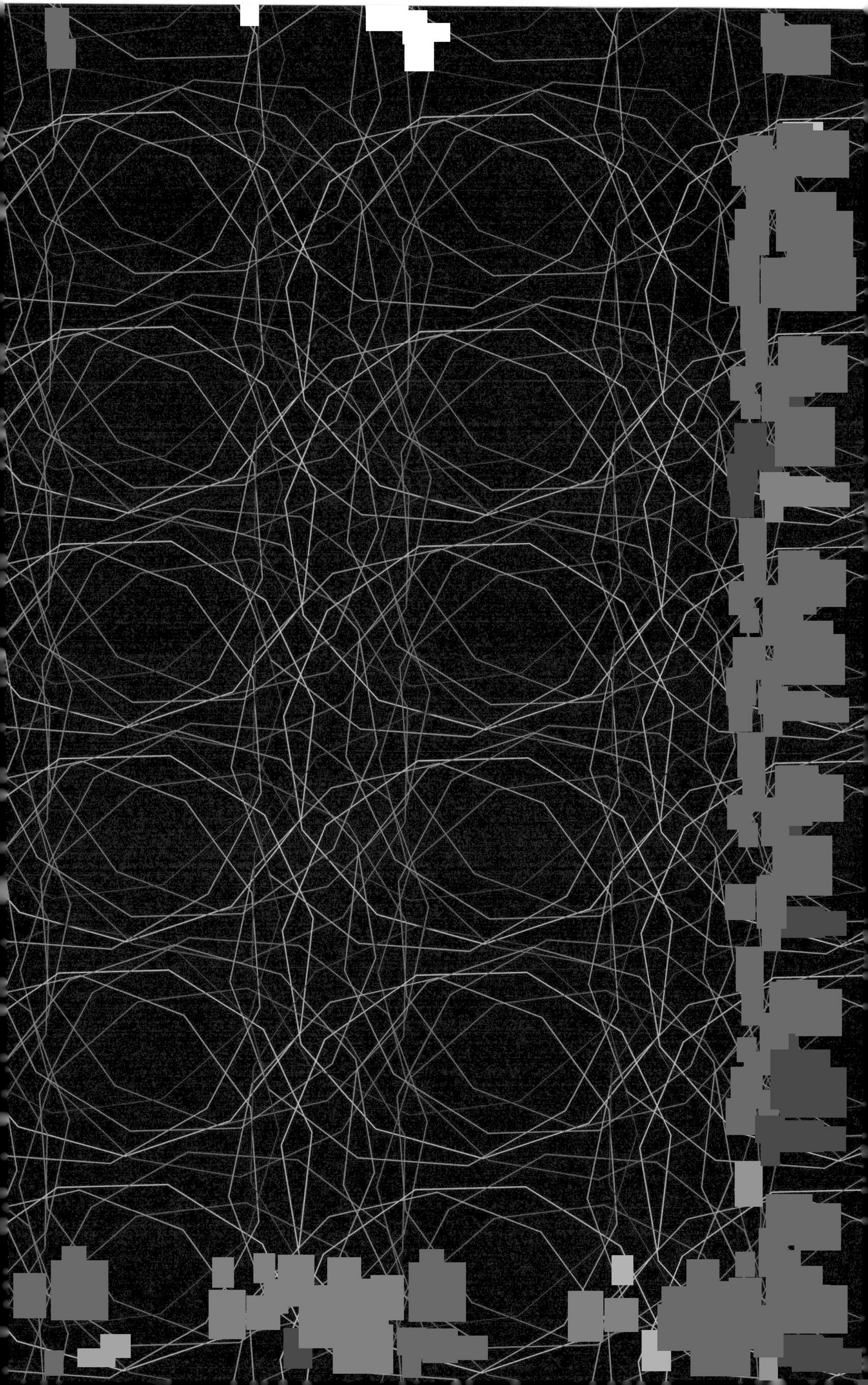

Generating a Geometric Pattern Workflow #5

Observe the geometric pattern, and analyze it to distinguish its constituent repeating motif.

The Motif

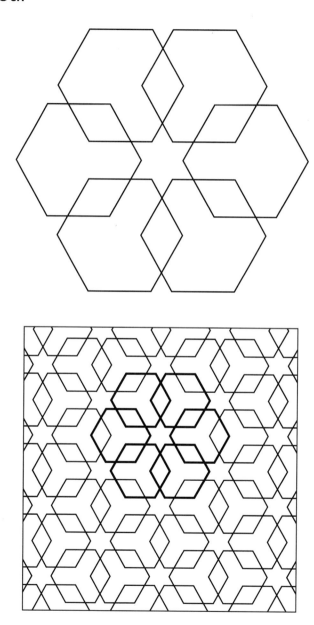

Analyzing the Constructive Elements

Step 1: We observe that there is an equal-sided hexagon with six repetitions rotated around the origin.

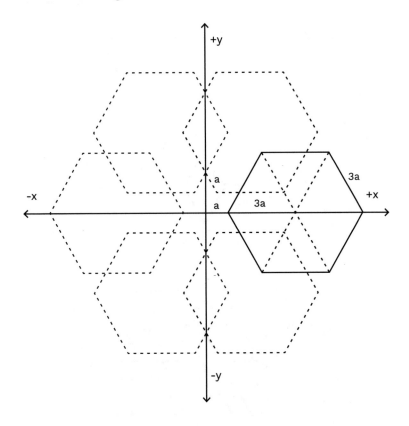

Generating the Motif

```
//scale factor
let a = 20;
function setup() {
  createCanvas(400, 400);
  angleMode(DEGREES);
  noLoop();
  noFill();
}
function draw() {
  push();
    translate(width*0.5,height*0.5);
    for(let i=0;i<6;i++){
      push();
        rotate(60*i);
        translate(4*a,0);
        beginShape();
        //hexagon
        for(let k=0;k<6;k++){
          let x = 3 * a * cos(k*60);
          let y = 3 * a * sin(k*60);
          vertex(x,y);
        }
        endShape(CLOSE);
      pop();
    }
  pop();
}
```

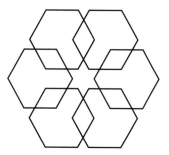

Analyzing the Tessellation

Step 2: We need to calculate the dx and dy values in the placement.

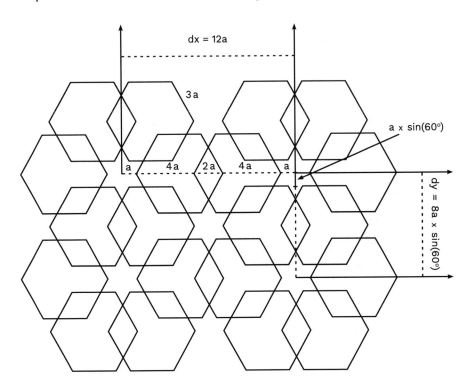

Tessellation Code

```
class Motif {
  constructor(a) {
    this.a = a;
  }

  display() {
    for(let i=0;i<6;i++){
      push();
        rotate(60*i);
        translate(4*this.a,0);
        beginShape();
        //hexagon
        for(let k=0;k<6;k++){
          let x = 3 * this.a * cos(k*60);
          let y = 3 * this.a * sin(k*60);
          vertex(x,y);
        }
        endShape(CLOSE);
      pop();
    }

  }
}
//scale factor
let a = 18;

let motif = new Motif(a);
let nRow;
let nCol;
let dx,dy;

function setup() {
  createCanvas(800, 800);
  angleMode(DEGREES);
  noFill();
  noLoop();

  dx = 12*a;
  dy = 8*a*sin(60);
  //approximate the nRow and nCol values
  nRow = ceil(height / dy);
  nCol = 1+ceil(width / dx);
}
```

Tessellation Code

```
function draw() {
  push();
    for (let r = 0; r < nRow; r++) {
      for (let c = 0; c < nCol; c++) {
        push();
          translate(dx*c,dy*r);
          motif.display();
        pop();
      }
    }
  pop();
}
```

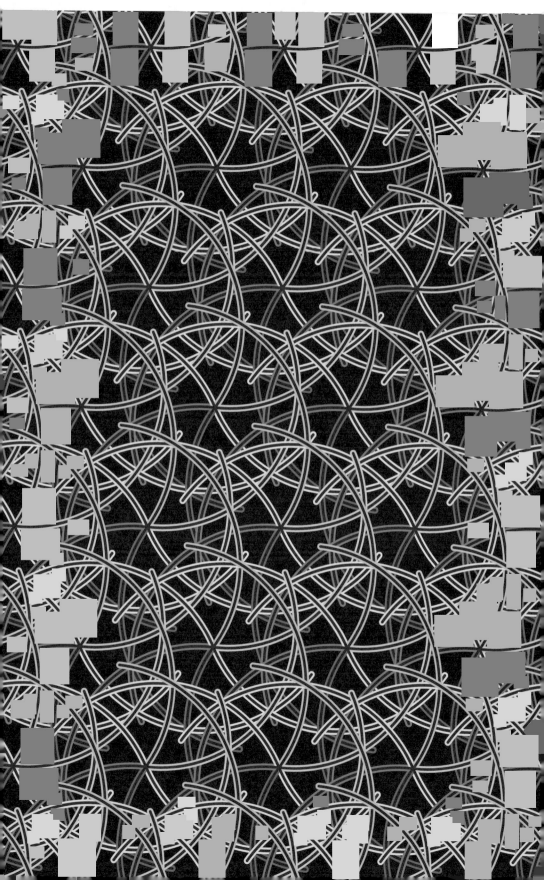

Generating a Geometric Pattern Workflow #6

Observe the geometric pattern, and analyze it to distinguish its constituent repeating motif.

The Motif

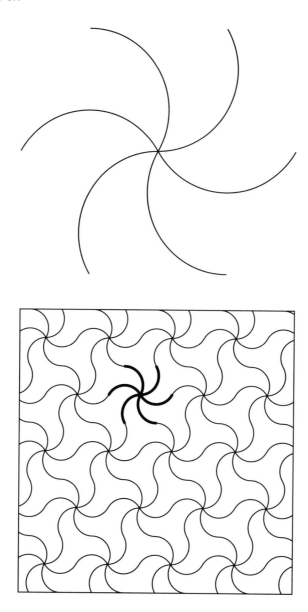

Analyzing the Constructive Elements

Step 1: Let's draw an equilateral hexagon around the motif.

Each arc is one-third of a complete circle.

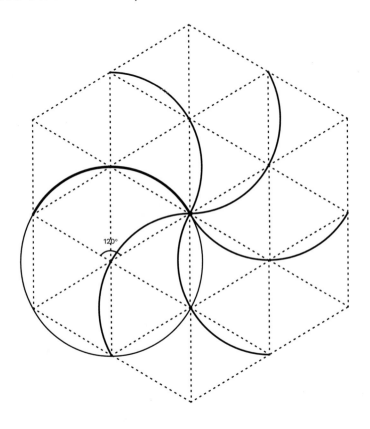

Step 2: In order to draw a section of a circle, we need to use the arc function of p5.js.

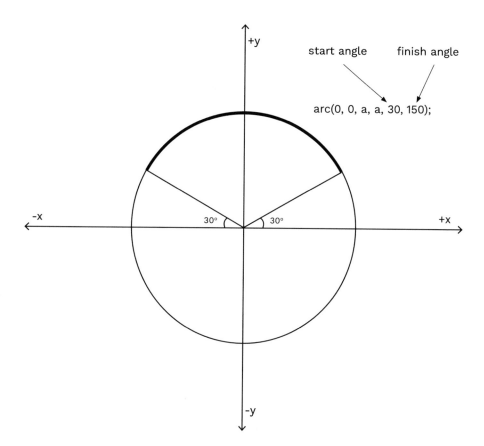

Step 3: Let's draw six of these arcs with their transformed positions and rotations.

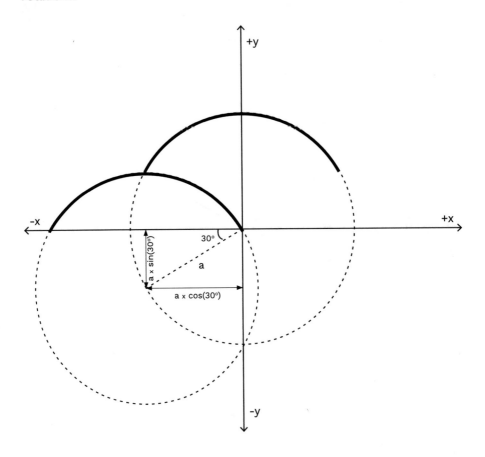

Step 4: Each arc is drawn in every 60° rotations.

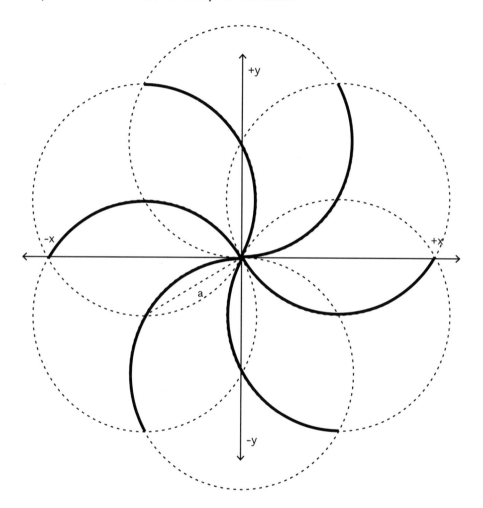

Generating the Motif

```
//scale factor
let a = 180;

function setup() {
    createCanvas(400, 400);
    angleMode(DEGREES);
    noFill();
    noLoop();
}
function draw() {
    let x,y;
    push();
        translate(width*0.5, height*0.5);
        for (let i = 0; i < 6; i++) {
            push();
                rotate(60 * i);
                translate(cos(30) * a * 0.5, -sin(30) * a * 0.5);
                arc(0, 0, a, a, 30, 150);
            pop();
        }
    pop();
}
```

Analyzing the Tessellation

Step 5: We need to calculate the dx, dy, and doff values in the placement.

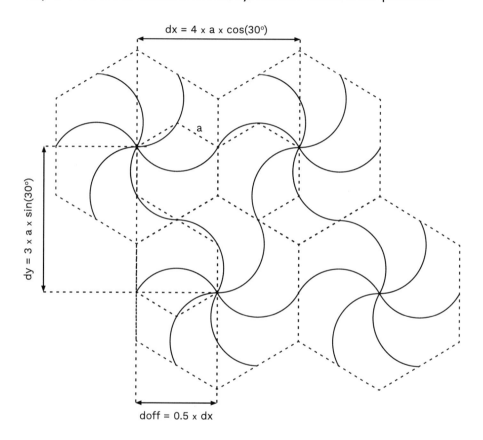

Tessellation Code

```
// Motif class
class Motif {
  constructor(a) {
    this.radius = a;
  }
  display() {

    for (let i = 0; i < 6; i++) {
      push();
        rotate(60 * i);
        translate(cos(30) * this.radius, -sin(30) * this.radius);
        arc(0, 0, this.radius * 2, this.radius * 2, 30, 150);
      pop();
    }
  }
}

//scale factor
let radius = 20;
let nRow;
let nCol;
let motif = new Motif(radius);
let dx,dy,doff;

function setup() {
  createCanvas(800, 800);
  angleMode(DEGREES);
  noLoop();
  noFill();

  dx = 4 * radius * cos(30);
  dy = 3 * radius;
  doff = 0.5 * dx;

  //approximate the nRow and nCol values
  nRow = ceil(height / dy);
  nCol = ceil(width / dx);

}
```

```
function draw() {
  for (let r = 0; r < nRow; r++) {
    for (let c = 0; c < nCol; c++) {
      push();
        translate(c*dx,  r*dy);
        if(r%2){
          //rows 0,2,4,6
          translate(doff, 0);
        }
        motif.display();
      pop();
    }
  }
}
```

Generating a Geometric Pattern Workflow #7

Observe the geometric pattern, and analyze it to distinguish its constituent repeating motif.

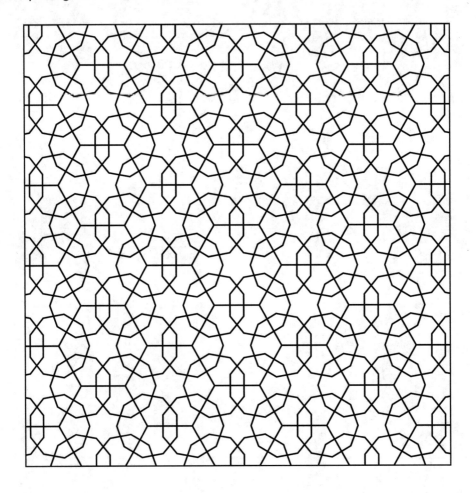

The Motif

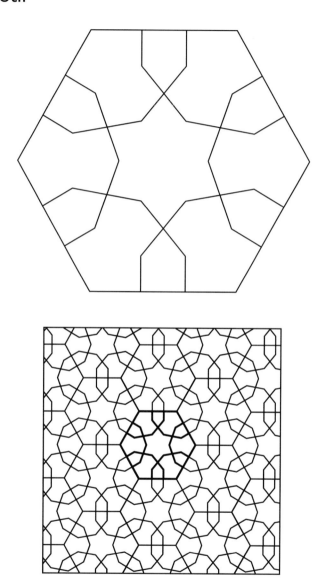

Analyzing the Constructive Elements

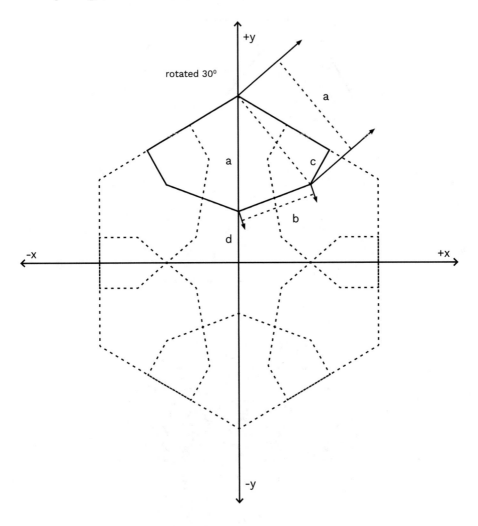

Calculating Angles and Vertex Points

Step 1: Let's find the vertex points of the constructive element.

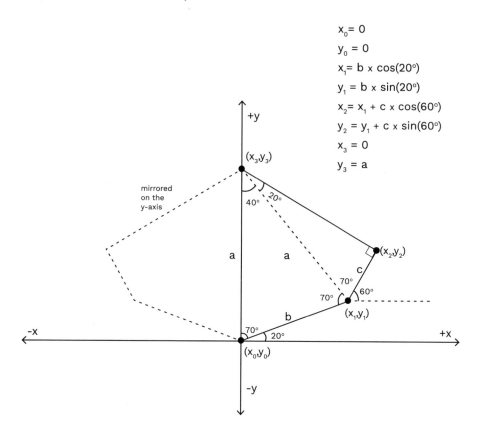

$$x_0 = 0$$
$$y_0 = 0$$
$$x_1 = b \times \cos(20°)$$
$$y_1 = b \times \sin(20°)$$
$$x_2 = x_1 + c \times \cos(60°)$$
$$y_2 = y_1 + c \times \sin(60°)$$
$$x_3 = 0$$
$$y_3 = a$$

Generating the Shape

```
//scale factor
let a = 120;
let b,c;
function setup() {
    createCanvas(400, 400);
    angleMode(DEGREES);
    noFill();
    noloop();
}

function draw() {
    push();
        translate(width * 0.5, height * 0.5);
        b = a*sin(40)/sin(70);
        c = b*0.5;

        let x0, y0, x1, y1, x2, y2, x3, y3;
        x0 = 0;
        y0 = 0;
        x1 = b * cos(20);
        y1 = b * sin(20);
        x2 = x1 + c * cos(60);
        y2 = y1 + c * sin(60);
        x3 = 0;
        y3 = a;

        beginShape();
        vertex(x0, -y0);
        vertex(x1, -y1);
        vertex(x2, -y2);
        vertex(x3, -y3);
        endShape();
        //mirrored on y axis
        beginShape();
        vertex(-x0, -y0);
        vertex(-x1, -y1);
        vertex(-x2, -y2);
        vertex(-x3, -y3);
        endShape();
    pop();
}
```

Analyzing the y-Offset of the Shape

Step 2: Let's find the vertex points of the constructive element.

$$\frac{d}{\sin(40°)} = \frac{c}{\sin(30°)}$$

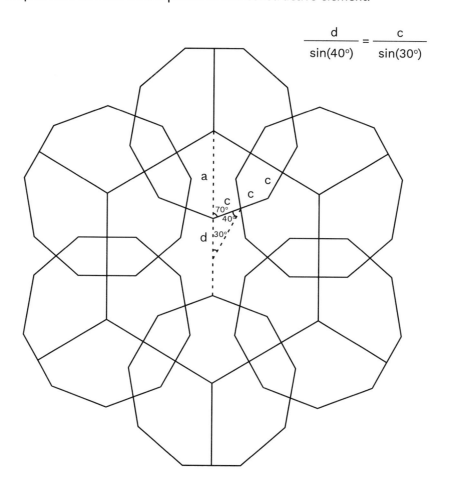

Generating the Motif

```
//scale factor
let a = 120;
let b, c, d;
function setup() {
    createCanvas(400, 400);
    angleMode(DEGREES);
    noFill();
    noLoop();
}

function draw() {
    push();
        translate(width * 0.5, height * 0.5);
        rotate(30); //align to the original shape

        b = (a * sin(40)) / sin(70);
        c = b * 0.5;
        d = (c * sin(40))/sin(30);
        let x0, y0, x1, y1, x2, y2, x3, y3;
        x0 = 0;
        y0 = 0;
        x1 = b * cos(20);
        y1 = b * sin(20);
        x2 = x1 + c * cos(60);
        y2 = y1 + c * sin(60);
        x3 = 0;
        y3 = a;

        for (let i = 0; i < 6; i++) {
            push();
                rotate(i * 60);
                translate(0, -d);
                beginShape();
                vertex(x0, -y0);
                vertex(x1, -y1);
                vertex(x2, -y2);
                vertex(x3, -y3);
                endShape();
```

```
        //mirrored on y axis
        beginShape();
        vertex(-x0, -y0);
        vertex(-x1, -y1);
        vertex(-x2, -y2);
        vertex(-x3, -y3);
        endShape();
      pop();
    }
  pop();
}
```

Analyzing the Tessellation

Step 3: We need to calculate the dx, dy, and doff values in the placement.

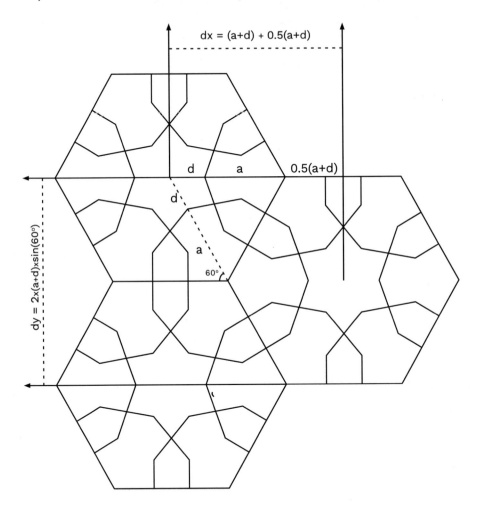

Tessellation Code

```
//Motif class
class Motif {
    constructor(a) {
        this.a = a;
    }

    display() {
        let a = this.a;
        let b, c, d;

        b = (a * sin(40)) / sin(70);
        c = b * 0.5;
        d = (c * sin(40)) / sin(30);

        let x0, y0, x1, y1, x2, y2, x3, y3;
        x0 = 0;
        y0 = 0;
        x1 = b * cos(20);
        y1 = b * sin(20);
        x2 = x1 + c * cos(60);
        y2 = y1 + c * sin(60);
        x3 = 0;
        y3 = a;

        push();
            rotate(30);
            for (let i = 0; i < 6; i++) {
                push();
                    rotate(i * 60);
                    translate(0, -d);
                    beginShape();
                    vertex(x0, -y0);
                    vertex(x1, -y1);
                    vertex(x2, -y2);
                    vertex(x3, -y3);
                    endShape();
                    //mirrored on y axis
                    beginShape();
                    vertex(-x0, -y0);
                    vertex(-x1, -y1);
                    vertex(-x2, -y2);
                    vertex(-x3, -y3);
                    endShape();
                pop();
            }
        pop();
    }
}
```

```
//scale factor
let a = 48;
let motif = new Motif(a);

let dx, dy, doff;
let nRow;
let nCol;

function setup() {
    createCanvas(800, 800);
    angleMode(DEGREES);
    noFill();
    noLoop();

    let b, c, d;
    b = (a * sin(40)) / sin(70);
    c = b * 0.5;
    d = (c * sin(40)) / sin(30);

    dx = 1.5 * (a+d);
    dy = 2 * (a+d) * sin(60);
    doff = 0.5 * dy;

    //approximate the nRow and nCol values
    nRow = 1 + ceil(height / dy);
    nCol = 1 + ceil(width / dx);
}

function draw() {
    for (let c = 0; c < nCol; c++) {
        for (let r = 0; r < nRow; r++) {
            push();
                if (c % 2 == 1){
                //columns 1,3,5,7
                translate(0, doff);
                }
                translate(dx * c, dy * r);
                motif.display();
            pop();
        }
    }
}
```

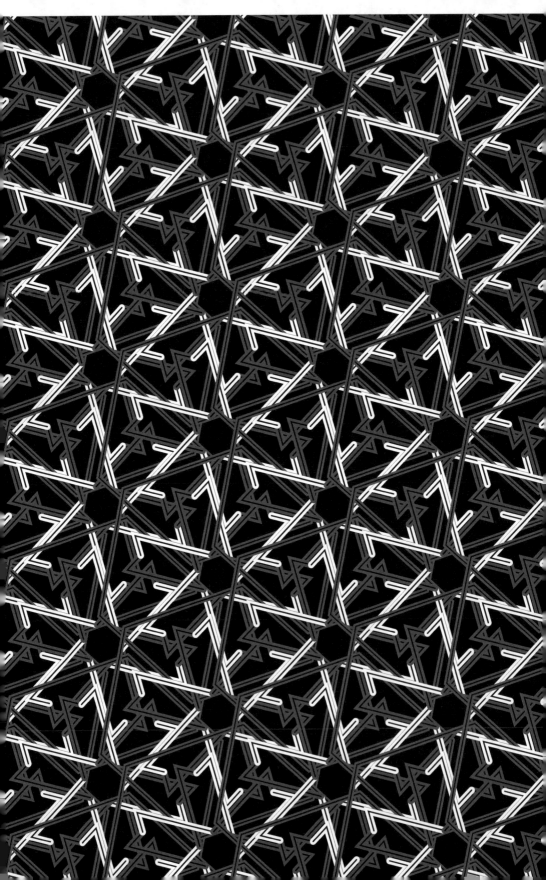

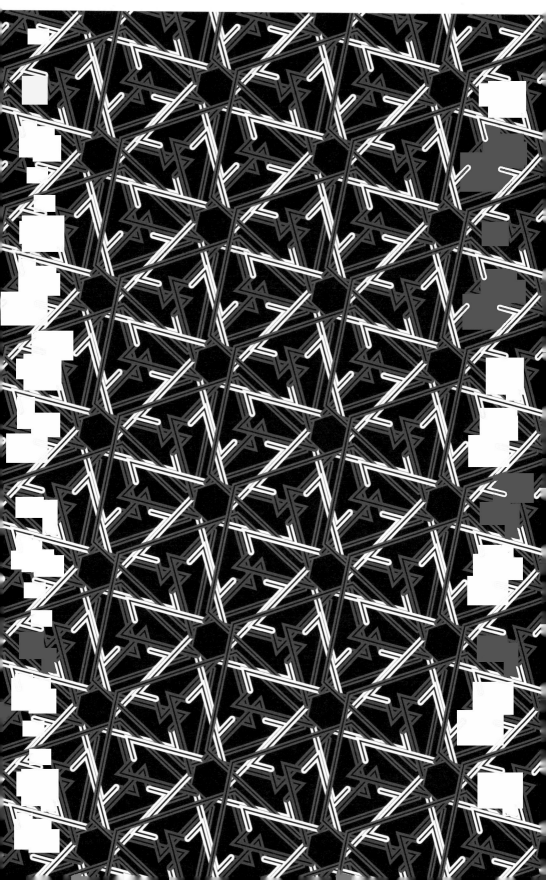

Generating a Geometric Pattern Workflow #8

Observe the geometric pattern, and analyze it to distinguish its constituent repeating motif.

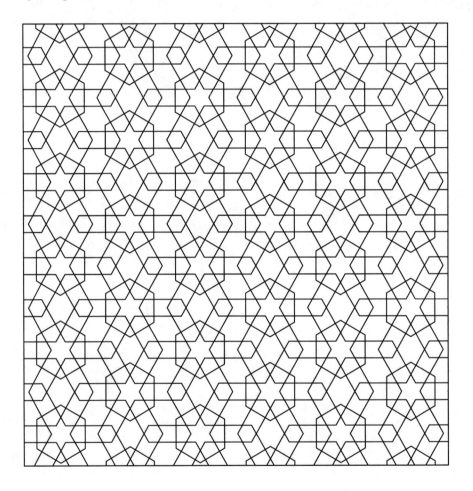

The Motif

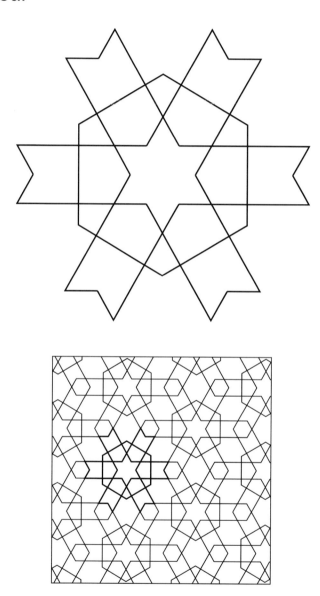

Analyzing the Constructive Elements

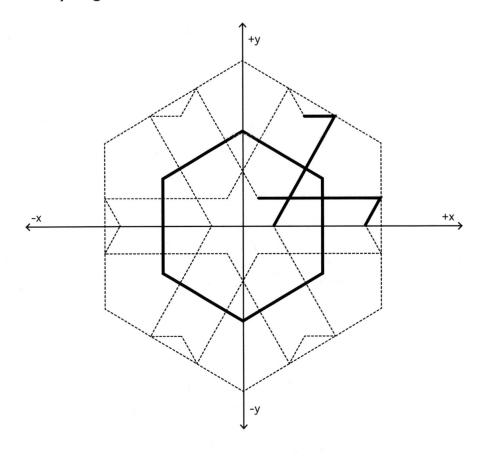

Calculating Angles and Vertex Points

Step 1: Let's find the vertex points of the constructive element.

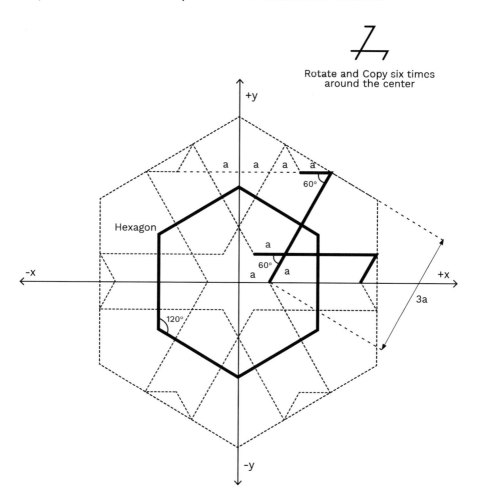

Rotate and Copy six times
around the center

Step 2: First, we will generate the shape as follows.

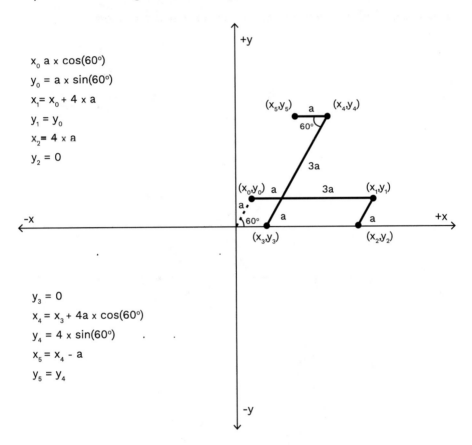

x_0 a x cos(60°)

$y_0 = a \times \sin(60°)$

$x_1 = x_0 + 4 \times a$

$y_1 = y_0$

$x_2 = 4 \times a$

$y_2 = 0$

$y_3 = 0$

$x_4 = x_3 + 4a \times \cos(60°)$

$y_4 = 4 \times \sin(60°)$

$x_5 = x_4 - a$

$y_5 = y_4$

Step 3: Now, we will draw our attention to the hexagon.

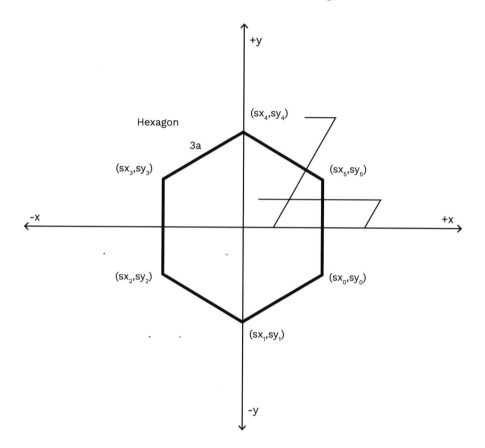

Generating the Motif

```
//scale factor
let a = 40;

function setup() {
    createCanvas(400, 400);
    angleMode(DEGREES);
    noFill();
    noLoop();
}

function draw() {
    let x0, y0, x1, y1, x2, y2, x3, y3, x4, y4, x5, y5;
    push();
        translate(width * 0.5, height * 0.5);
        //vertex points
        x0 = a * cos(60);
        y0 = -a * sin(60);
        x1 = x0 + 4 * a;
        y1 = y0;
        x2 = 4 * a;
        y2 = 0;
        x3 = a;
        y3 = 0;
        x4 = x3 + 4 * a * cos(60);
        y4 = -4 * a * sin(60);
        x5 = x4 - a;
        y5 = y4;
        for (let i = 0; i < 6; i++) {
            push();
                rotate(60 * i);
                beginShape();
                vertex(x0, y0);
                vertex(x1, y1);
                vertex(x2, y2);
                endShape();
                beginShape();
                vertex(x3, y3);
                vertex(x4, y4);
                vertex(x5, y5);
                endShape();
            pop();
        }
}
```

```
    //hexagon
    //needs to be rotated 90 degrees to adjust the orientation
    rotate(90);
    let angle = 360 / 6;
    beginShape();
    for (let ang = 0; ang < 360; ang += angle) {
        let sx = cos(ang) * 3 * a;
        let sy = sin(ang) * 3 * a;
        vertex(sx, sy);
    }
    endShape(CLOSE);
  pop();
}
```

Analyzing the Tessellation

Step 4: We need to calculate the dx, dy, and doff values in the placement.

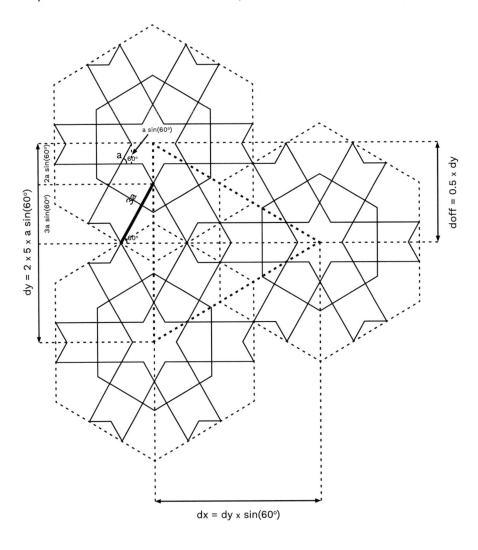

$dx = dy \times \sin(60°)$

Tessellation Code

```
// Motif class
class Motif {
    constructor(a) {
        this.a = a;
    }

    display() {
        let x0, y0, x1, y1, x2, y2, x3, y3, x4, y4, x5, y5;
        x0 = a * cos(60);
        y0 = -a * sin(60);
        x1 = x0 + 4 * a;
        y1 = y0;
        x2 = 4 * a;
        y2 = 0;
        x3 = a;
        y3 = 0;
        x4 = x3 + 4 * a * cos(60);
        y4 = -4 * a * sin(60);
        x5 = x4 - a;
        y5 = y4;
        for (let i = 0; i < 6; i++) {
            push();
                rotate(60 * i);
                beginShape();
                vertex(x0, y0);
                vertex(x1, y1);
                vertex(x2, y2);
                endShape();
                beginShape();
                vertex(x3, y3);
                vertex(x4, y4);
                vertex(x5, y5);
                endShape();
            pop();
        }
    }
```

```
        //hexagon
        //needs to be rotated 90 degrees to adjust the orientation
        rotate(90);
        let angle = 360 / 6;
        beginShape();
        for (let ang = 0; ang < 360; ang += angle) {
            let sx = cos(ang) * 3 * a;
            let sy = sin(ang) * 3 * a;
            vertex(sx, sy);
        }
        endShape(CLOSE);
    }
}

//scale factor
let a = 16;
let motif = new Motif(a);
let nRow;
let nCol;
let dx, dy, doff;

function setup() {
    createCanvas(800, 800);
    angleMode(DEGREES);
    noFill();
    noLoop();

    dy = 2 * 5 * a * sin(60);
    dx = dy * sin(60);
    doff = dy * 0.5;

    //approximate the nRow and nCol values
    nRow = 1 + ceil(height / dy);
    nCol = 1 + ceil(width / dx);
}
```

```
function draw() {
    for (let c = 0; c < nCol; c++) {
        for (let r = 0; r < nRow; r++) {
            push();
                if (c % 2 == 1) {
                    //columns 1,3,5,7
                    translate(0, doff);
                }
                translate(dx * c, dy * r);
                motif.display();
            pop();
        }
    }
}
```

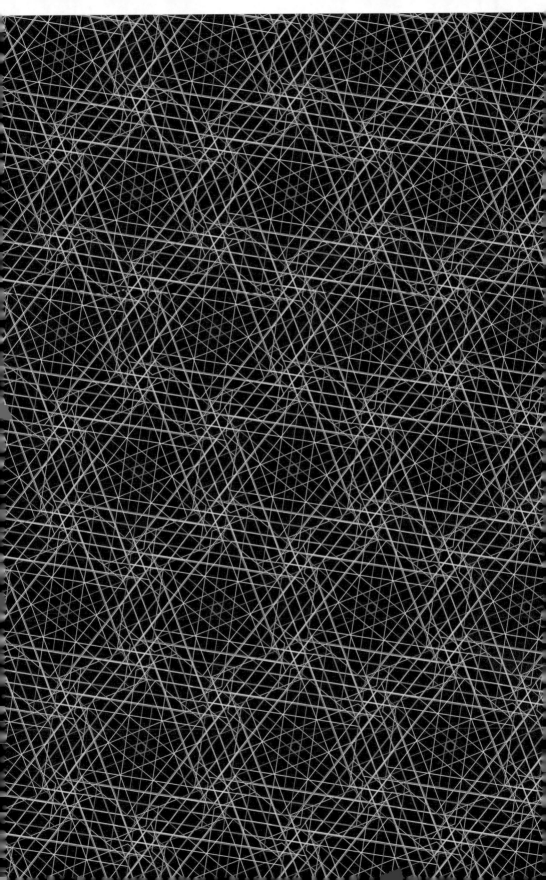

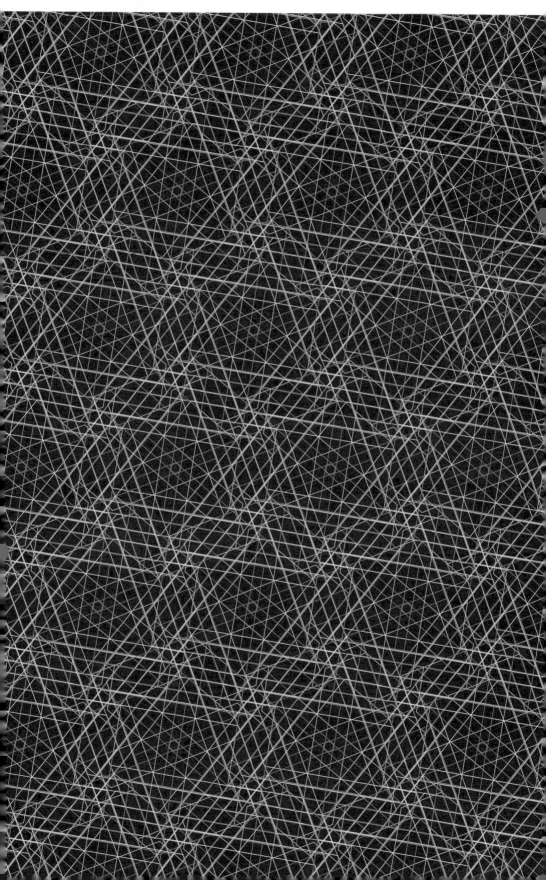

Generating a Geometric Pattern Workflow #9

Observe the geometric pattern, and analyze it to distinguish its constituent repeating motif.

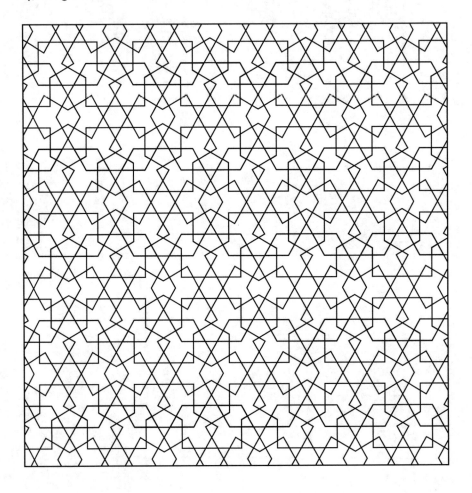

The Motif

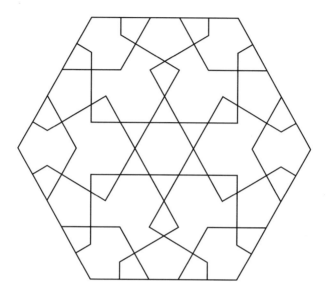

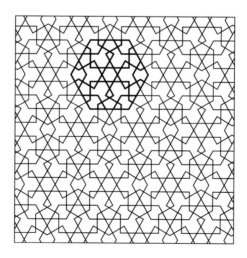

Analyzing the Constructive Elements

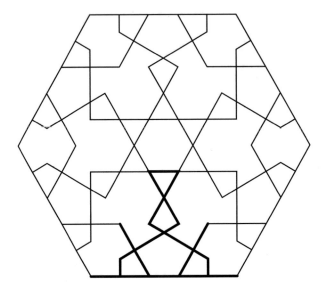

Calculating Angles and Vertex Points

Step 1: Let's find the vertex points of the constructive element.

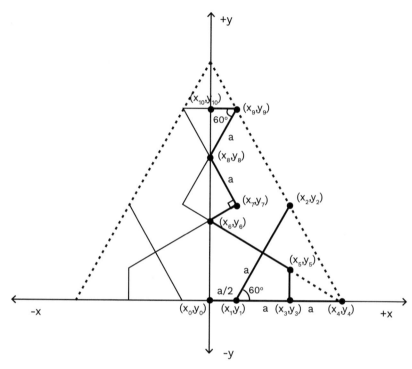

$x_0 = 0$

$y_0 = 0$

$x_1 = 0.5 \times a$

$y_1 = 0$

$x_2 = 2 \times a \times \cos(60°) + 0.5 \times a$

$y_2 = 2 \times a \times \sin(60°)$

$x_3 = 1.5 \times a$

$y_3 = 0$

$x_4 = 2.5 \times a$

$y_4 = 0$

$x_5 = 1.5 \times a$

$y_5 = a/\tan(60°)$

$x_6 = 0$

$y_6 = 2.5 \times a/\tan(60°)$

$x_7 = a \times \sin(30°)$

$y_7 = 2 \times a \times \sin(60°)$

$x_8 = 0$

$y_8 = 3 \times a \times \sin(60°)$

$x_9 = a \times \sin(30°)$

$y_9 = 4 \times a \times \sin(60°)$

$x_{10} = 0$

$y_{10} = 4 \times a \times \sin(60°)$

Step 2: The motif needs to be translated down and rotated 60 degrees each time copied.

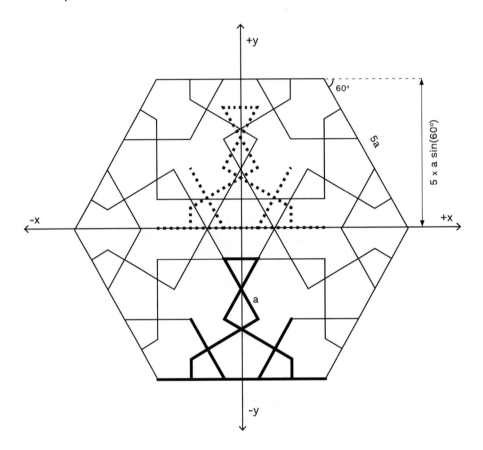

Generating the Motif

```
//scale factor
let a = 30;

function setup() {
  createCanvas(400, 400);
  angleMode(DEGREES);
  noLoop();
  noFill();
}

function draw() {
    let x0,y0,x1,y1,x2,y2,x3,y3,x4,y4,x5,y5,x6,y6,x7,y7,x8,y8,x9,y9,x10,y10;
    push();
      translate(width*0.5, height*0.5);
      x0 = 0;
      y0 = 0;
      x1 = 0.5 * a;
      y1 = 0;
      x2 = 2 * a * cos(60) + 0.5 * a;
      y2 = -(2 * a * sin(60));
      x3 = 1.5 * a;
      y3 = 0;
      x4 = 2.5 * a;
      y4 = 0;
      x5 = 1.5 * a;
      y5 = -a / tan(60);
      x6 = 0;
      y6 = (-2.5 * a) / tan(60);
      x7 = a * sin(30);
      y7 = -(2 * a * sin(60));
      x8 = 0;
      y8 = -(3 * a * sin(60));
      x9 = a * sin(30);
      y9 = -(4 * a * sin(60));
      x10 = 0;
      y10 = -(4 * a * sin(60));

      for (let i = 0; i < 6; i++) {
        push();
          rotate(i * 60);
          translate(0, 5 * a * sin(60));

          beginShape();
          vertex(x0, y0);
          vertex(x1, y1);
```

```
        vertex(x2, y2);
        endShape();

        beginShape();
        vertex(x1, y1);
        vertex(x4, y4);
        endShape();

        beginShape();
        vertex(x3, y3);
        vertex(x5, y5);
        vertex(x6, y6);
        vertex(x7, y7);
        vertex(x8, y8);
        vertex(x9, y9);
        vertex(x10, y10);
        endShape();

        //mirror on y axis
        beginShape();
        vertex(-x0, y0);
        vertex(-x1, y1);
        vertex(-x2, y2);
        endShape();

        beginShape();
        vertex(-x1, y1);
        vertex(-x4, y4);
        endShape();

        beginShape();
        vertex(-x3, y3);
        vertex(-x5, y5);
        vertex(-x6, y6);
        vertex(-x7, y7);
        vertex(-x8, y8);
        vertex(-x9, y9);
        vertex(-x10, y10);
        endShape();

      pop();
    }
  pop();
}
```

Analyzing the Tessellation

Step 3: We need to calculate the dx, dy, and doff values in the placement.

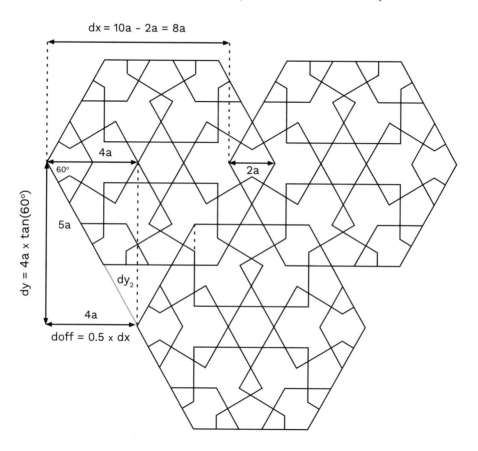

Tessellation Code

```
class Motif {
  constructor(a) {
    this.a = a;
  }
  display() {

    let x0,y0,x1,y1,x2,y2,x3,y3,x4,y4,x5,y5,x6,y6,x7,y7,x8,y8,x9,y9,x10,y10;
    x0 = 0;
    y0 = 0;
    x1 = 0.5 * this.a;
    y1 = 0;
    x2 = 2 * this.a * cos(60) + 0.5 * this.a;
    y2 = -(2 * this.a * sin(60));
    x3 = 1.5 * this.a;
    y3 = 0;
    x4 = 2.5 * this.a;
    y4 = 0;
    x5 = 1.5 * this.a;
    y5 = -this.a / tan(60);
    x6 = 0;
    y6 = (-2.5 * this.a) / tan(60);
    x7 = this.a * sin(30);
    y7 = -(2 * this.a * sin(60));
    x8 = 0;
    y8 = -(3 * this.a * sin(60));
    x9 = this.a * sin(30);
    y9 = -(4 * this.a * sin(60));
    x10 = 0;
    y10 = -(4 * this.a * sin(60));
    for (let i = 0; i < 6; i++) {
      push();
        rotate(i * 60);
        translate(0, 5 * this.a * sin(60));

        beginShape();
        vertex(x0, y0);
        vertex(x1, y1);
        vertex(x2, y2);
        endShape();

        beginShape();
        vertex(x1, y1);
        vertex(x4, y4);
        endShape();
```

```
        beginShape();
        vertex(x3, y3);
        vertex(x5, y5);
        vertex(x6, y6);
        vertex(x7, y7);
        vertex(x8, y8);
        vertex(x9, y9);
        vertex(x10, y10);
        endShape();

        //mirror on y axis
        beginShape();
        vertex(-x0, y0);
        vertex(-x1, y1);
        vertex(-x2, y2);
        endShape();

        beginShape();
        vertex(-x1, y1);
        vertex(-x4, y4);
        endShape();

        beginShape();
        vertex(-x3, y3);
        vertex(-x5, y5);
        vertex(-x6, y6);
        vertex(-x7, y7);
        vertex(-x8, y8);
        vertex(-x9, y9);
        vertex(-x10, y10);
        endShape();

      pop();
    }
  }
}

//scale factor
let a = 16;
let motif = new Motif(a);
let nRow;
let nCol;
let dx, dy, doff;
```

```
function setup() {
  createCanvas(800, 800);
  angleMode(DEGREES);
  noFill();
  noLoop();

  dx = 8.0*a;
  dy = 4.0*a*tan(60);
  doff = 0.5*dx;

  //approximate the nRow and nCol values
  nRow = ceil(height / dy);
  nCol = ceil(width / dx);
}

function draw() {
  for (let r = 0; r < nRow; r++) {
    for (let c = 0; c < nCol; c++) {
      push();
        translate(c * dx, r * dy);
        if (r % 2 == 1) {
          //rows 1,3,5,7
          translate(doff,0);
        }
        motif.display();
      pop();
    }
  }
}
```

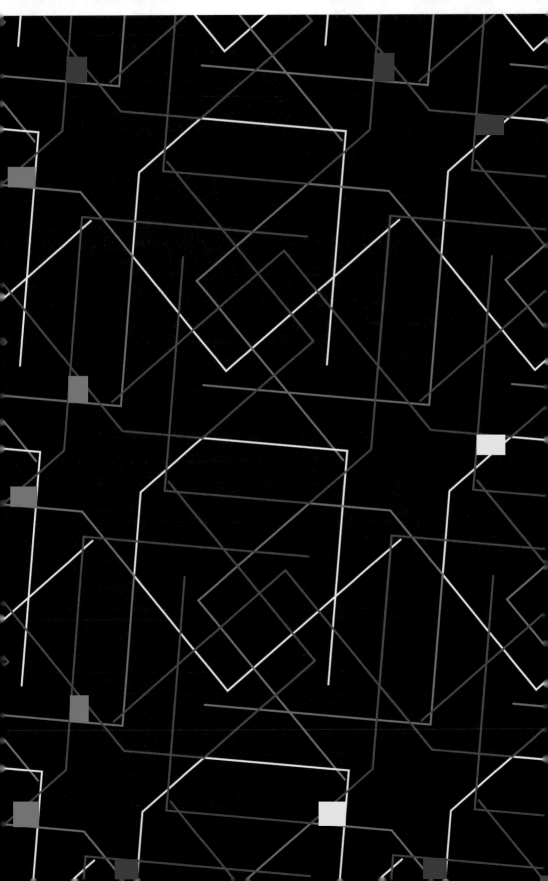

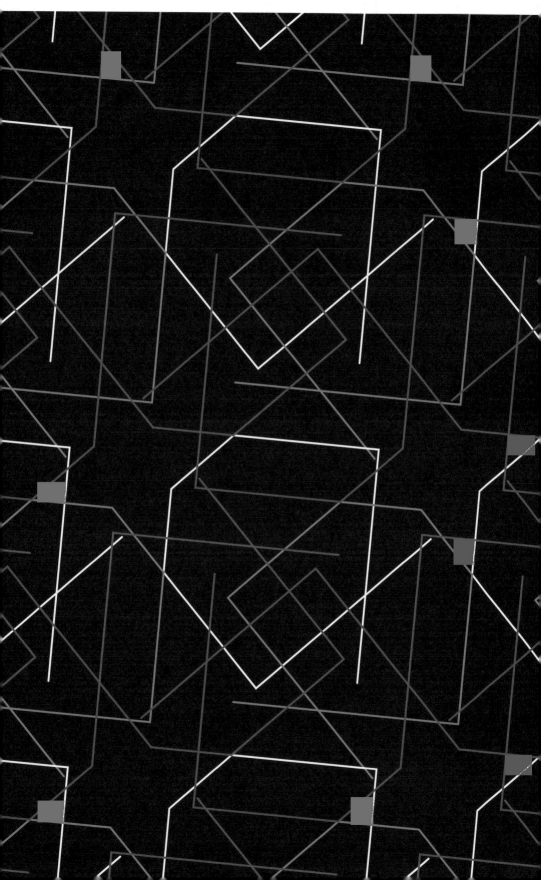

Generating a Geometric Pattern Workflow #10

Observe the geometric pattern, and analyze it to distinguish its constituent repeating motif.

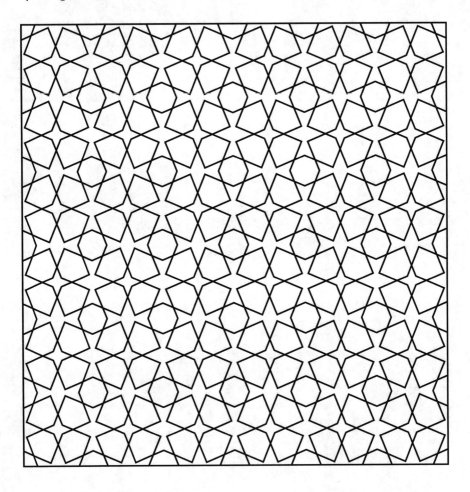

The Motif

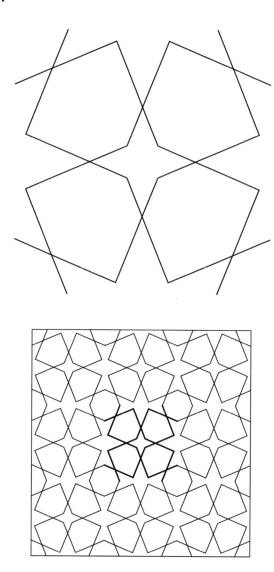

Analyzing the Constructive Elements

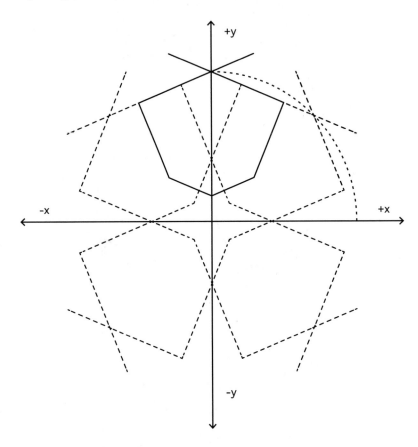

Calculating Angles and Vertex Points

Step 1: Let's find the vertex points of the constructive element.

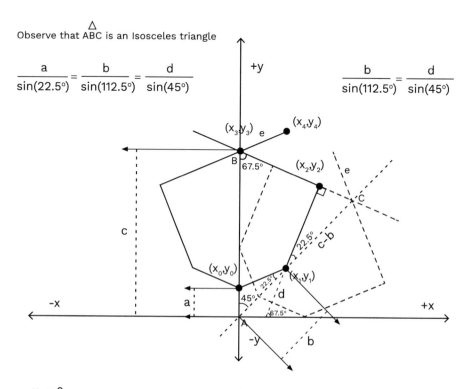

Observe that $\overset{\triangle}{ABC}$ is an Isosceles triangle

$$\frac{a}{\sin(22.5°)} = \frac{b}{\sin(112.5°)} = \frac{d}{\sin(45°)}$$

$$\frac{b}{\sin(112.5°)} = \frac{d}{\sin(45°)}$$

$x_0 = 0$

$y_0 = a$

$x_1 = b \times \cos(45°)$

$y_1 = b \times \sin(45°)$

$x_2 = (d + (c-b) \times \cos(22.5°)) \times \cos(67.5°) + a$

$y_2 = (d + (c-b) \times \cos(22.5°)) \times \sin(67.5°)$

$x_3 = 0$

$y_3 = c$

$x_4 = e \times \cos(22.5°)$

$y_4 = e \times \sin(22.5°) + c$

Generating the Motif

```
//scale factor
let a = 25;
let b, c, d, e;

function setup() {
    createCanvas(400, 400);
    angleMode(DEGREES);
    noLoop();
    noFill();
    a = 25;
    b = (a * sin(112.5)) / sin(22.5);
    c = 6 * a;
    d = (a * sin(45)) / sin(22.5);
    e = ((c - b) * sin(22.5)) / sin(45);
}

function draw() {
    let x0, y0, x1, y1, x2, y2, x3, y3, x4, y4;
    push();
        translate(width * 0.5, height * 0.5);
        rotate(45);
        for (let n = 0; n < 4; n++) {
            push();
                rotate(90 * n);
                beginShape();
                x0 = 0;
                y0 = -a;
                vertex(x0, y0);
                x1 = b * cos(45);
                y1 = -b * sin(45);
                vertex(x1, y1);
                x2 = (d + (c - b) * cos(22.5)) * cos(67.5) + a;
                y2 = -1 * ((d + (c - b) * cos(22.5)) * sin(67.5));
                vertex(x2, y2);
                x3 = 0;
                y3 = -c;
                vertex(x3, y3);
                x4 = e * cos(22.5);
                y4 = -1 * (e * sin(22.5) + c);
                vertex(x4, y4);
                endShape();
```

```
//mirroring
push();
    scale(-1, 1);
    beginShape();
    x0 = 0;
    y0 = -a;
    vertex(x0, y0);
    x1 = b * cos(45);
    y1 = -b * sin(45);
    vertex(x1, y1);
    x2 = (d + (c - b) * cos(22.5)) * cos(67.5) + a;
    y2 = -1 * ((d + (c - b) * cos(22.5)) * sin(67.5));
    vertex(x2, y2);
    x3 = 0;
    y3 = -c;
    vertex(x3, y3);
    x4 = e * cos(22.5);
    y4 = -1*(e * sin(22.5) + c);
    vertex(x4, y4);
    endShape();
    pop();
    pop();
    }
pop();
}
```

Analyzing the Tessellation

Step 2: We need to calculate the dx and dy values in the placement.

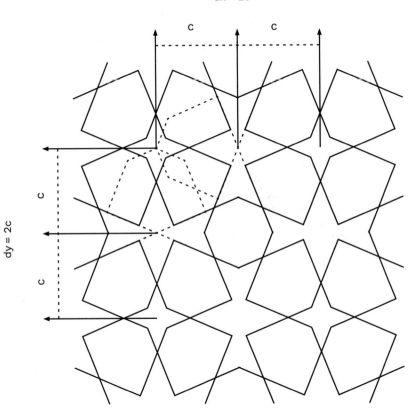

Tessellation Code

```
// Motif class
class Motif {
    constructor(a) {
        this.a = a;
    }

    display() {
        let b = (this.a * sin(112.5)) / sin(22.5);
        let c = this.a * 6;
        let d = (this.a * sin(45)) / sin(22.5);
        let e = ((c - b) * sin(22.5)) / sin(45);

        let x0, y0, x1, y1, x2, y2, x3, y3, x4, y4;
        rotate(45);
        for (let n = 0; n < 4; n++) {
            push();
                rotate(90 * n);
                beginShape();
                x0 = 0;
                y0 = -this.a;
                vertex(x0, y0);
                x1 = b * cos(45);
                y1 = -b * sin(45);
                vertex(x1, y1);
                x2 = (d + (c - b) * cos(22.5)) * cos(67.5) + this.a;
                y2 = -1 * ((d + (c - b) * cos(22.5)) * sin(67.5));
                vertex(x2, y2);
                x3 = 0;
                y3 = -c;
                vertex(x3, y3);
                x4 = e * cos(22.5);
                y4 = -1 * (e * sin(22.5) + c);
                vertex(x4, y4);
                endShape();
                //mirroring
                push();
                    scale(-1, 1);
                    beginShape();
                    x0 = 0;
                    y0 = -this.a;
                    vertex(x0, y0);
                    x1 = b * cos(45);
                    y1 = -b * sin(45);
                    vertex(x1, y1);
```

```
                x2 = (d + (c - b) * cos(22.5)) * cos(67.5) + this.a;
                y2 = -1 * ((d + (c - b) * cos(22.5)) * sin(67.5));
                vertex(x2, y2);
                x3 = 0;
                y3 = -c;
                vertex(x3, y3);
                x4 = e * cos(22.5);
                y4 = -1 * (e * sin(22.5) + c);
                vertex(x4, y4);
                endShape();
            pop();
        pop();
      }
    }
}

//scale factor
let a = 20;
let c;
let motif = new Motif(a);
let nRow;
let nCol;

function setup() {
    createCanvas(800, 800);
    angleMode(DEGREES);
    noLoop();
    noFill();

    c = 6 * a;

    //approximate the nRow and nCol values
    nRow = ceil(height / (2 * c));
    nCol = ceil(width / (2 * c));
}

function draw() {
    for (let r = 0; r < nRow; r++) {
        for (let k = 0; k < nCol; k++) {
            push();
                translate(k * c * 2, c * r * 2);
                motif.display();
            pop();
        }
    }
}
```

Generating a Geometric Pattern Workflow #11

Observe the geometric pattern, and analyze it to distinguish its constituent repeating motif.

The Motif

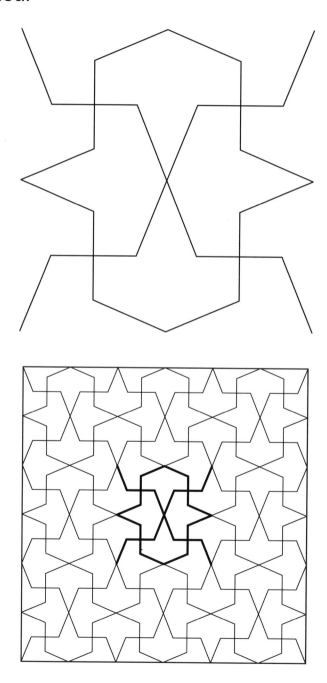

Analyzing the Constructive Elements

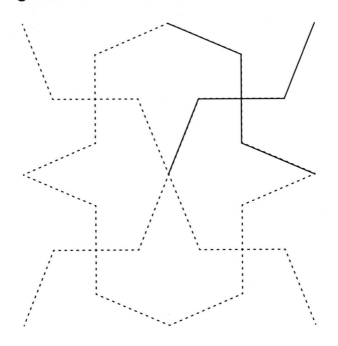

Calculating Angles and Vertex Points

Step 1: Let's find the vertex points of the constructive element.

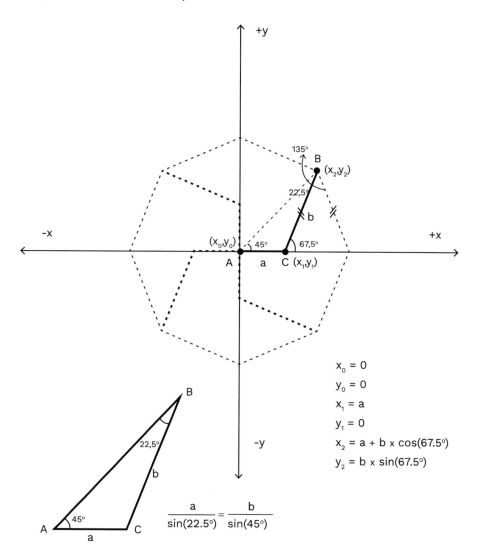

$x_0 = 0$

$y_0 = 0$

$x_1 = a$

$y_1 = 0$

$x_2 = a + b \times \cos(67.5°)$

$y_2 = b \times \sin(67.5°)$

$$\frac{a}{\sin(22.5°)} = \frac{b}{\sin(45°)}$$

Generating the Motif

```
//scale factor
let a = 40;
let b;
function setup() {
  createCanvas(400, 400);
  angleMode(DEGREES);
  noLoop();
  noFill();
  b = a * (sin(45) / sin(22.5));
}

function draw() {
  let x0,y0,x1,y1,x2,y2;

  push();
    translate(width*0.5,height*0.5);
    for(let i=0;i<4;i++){
      rotate(i*90);
      beginShape();
      x0 = 0;
      y0 = 0;
      x1 = a;
      y1 = 0;
      x2 = a + b * cos(67.5);
      y2 = b * sin(67.5);
      vertex(x0,-y0);
      vertex(x1,-y1);
      vertex(x2,-y2);
      endShape();
    }
  pop();
}
```

Step 2: We need to generate the complete motif with using the quadrant.

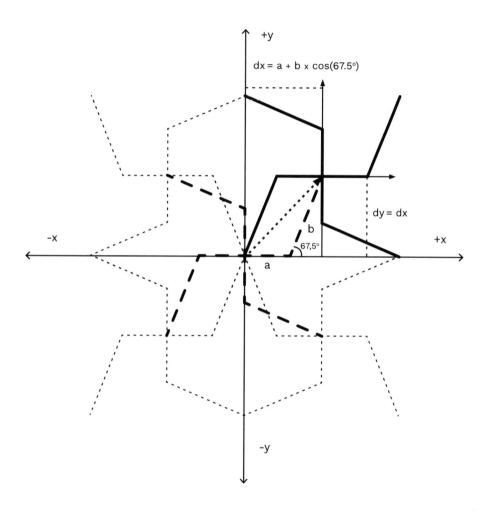

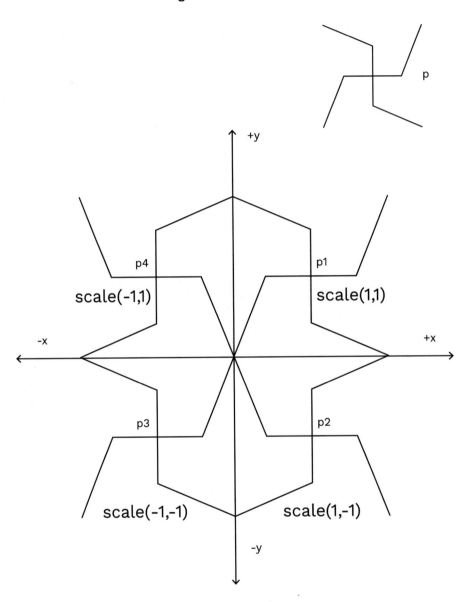

p2 is the mirror reflection of p1 with respect to the horizontal axis.

p3 is the mirror reflection of p1 with respect to the horizontal and vertical axes.

p4 is the mirror reflection of p1 with respect to the vertical axis.

We use the *scale* transformation function to create reflections.

Generating the Motif

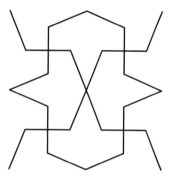

```
//scale factor
let a = 40;
let b;
function setup() {
  createCanvas(400, 400);
  angleMode(DEGREES);
  noLoop();
  noFill();
  b = a * (sin(45) / sin(22.5));
}

function draw() {
  let x0,y0,x1,y1,x2,y2;
  let dx = a + b * cos(67.5);
  let dy = dx;

  push();
    translate(width*0.5,height*0.5);
    for(let k=0;k<4;k++){
      push();
        //mirroring
        switch (k) {
          case 0:
            scale(1,1);
            break;
          case 1:
            scale(1,-1);
            break;
          case 2:
            scale(-1,-1);
            break;
          case 3:
            scale(-1,1);
            break;
          default:
            //
        }
```

Generating the Motif

```
translate(dx,-dy);
//quarter shape
for(let i=0;i<4;i++){
  rotate(i*90);
  beginShape();
  x0 = 0;
  y0 = 0;
  x1 = a;
  y1 = 0;
  x2 = a + b * cos(67.5);
  y2 = b * sin(67.5);
  vertex(x0,-y0);
  vertex(x1,-y1);
  vertex(x2,-y2);
  endShape();
  }
  pop();
}
pop();
}
```

Analyzing the Tessellation

Step 3: We need to calculate the dx and dy values in the placement.

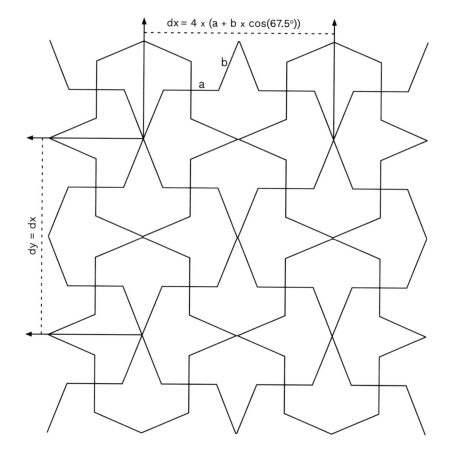

Tessellation Code

```
//Motif class
class Motif {
  constructor(a) {
    this.a = a;
  }

  display() {
    let x0,y0,x1,y1,x2,y2;
    let b = this.a * (sin(45) / sin(22.5));
    let dx = this.a + b * cos(67.5);
    let dy = dx;

    for(let k=0;k<4;k++){
      push();
        //mirroring
        switch (k) {
         case 0:
          scale(1,1);
          break;
         case 1:
          scale(1,-1);
          break;
         case 2:
          scale(-1,-1);
          break;
         case 3:
          scale(-1,1);
          break;
         default:
          //
        }
        translate(dx,-dy);
        //quarter shape
        for(let i=0;i<4;i++){
          rotate(i*90);
          beginShape();
          x0 = 0;
          y0 = 0;
          x1 = a;
          y1 = 0;
          x2 = a + b * cos(67.5);
          y2 = b * sin(67.5);
          vertex(x0,-y0);
          vertex(x1,-y1);
          vertex(x2,-y2);
          endShape();
        }
```

```
        pop();
      }
    }
  }

//scale factor
let a = 20;
let motif = new Motif(a);
let xOff,yOff;
let nRow;
let nCol;

function setup() {
  createCanvas(800, 800);
  angleMode(DEGREES);
  noLoop();
  noFill();

  let b = a * (sin(45) / sin(22.5));
  xOff = 4 * (a + b * cos(67.5));
  yOff = xOff;

  //approximate the nRow and nCol values
  nRow = 1+ceil(height / xOff);
  nCol = 1+ceil(width / yOff);
}

function draw() {
  push();
    for (let r = 0; r < nRow; r++) {
      for (let c = 0; c < nCol; c++) {
        push();
          translate(xOff * c,  yOff * r);
          motif.display();
        pop();
      }
    }
  pop();
}
```

Generating a Geometric Pattern Workflow #12

Observe the geometric pattern, and analyze it to distinguish its constituent repeating motif.

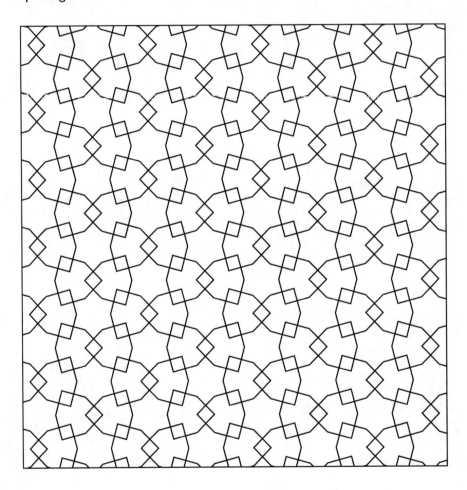

The Motif

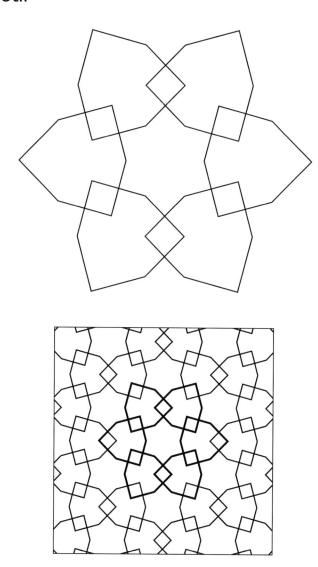

Analyzing the Constructive Elements

Step 1: Let's find the vertex points of the constructive element.

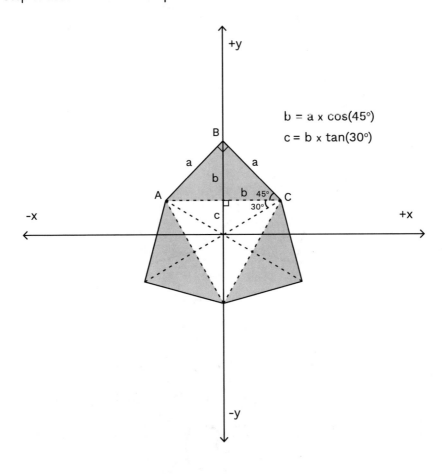

Step 2: Let's find the vertex points of the constructive triangle.

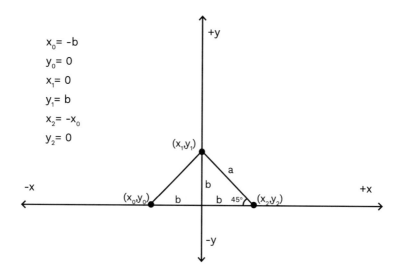

Step 3: The triangle shape needs to be copied, translated, and rotated three times around the origin.

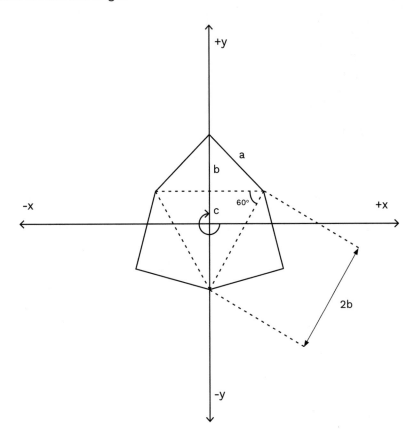

Step 4: Mini squares are formed in between the intersections of the basic shape. The size of the square depends on the shape's translation dimension. Here, we assume that the square size is equal to half of "a".

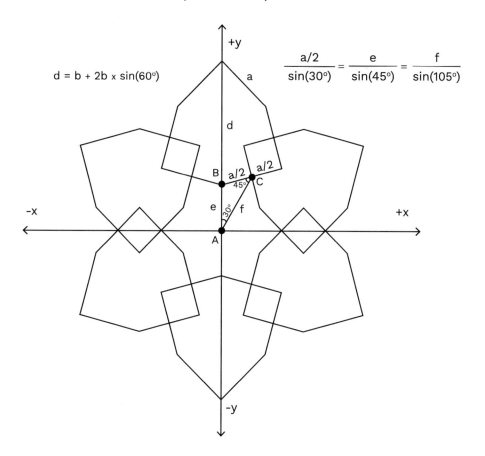

$d = b + 2b \times \sin(60°)$

$$\frac{a/2}{\sin(30°)} = \frac{e}{\sin(45°)} = \frac{f}{\sin(105°)}$$

Step 5: The motif needs a 90-degree rotation to finalize its placement.

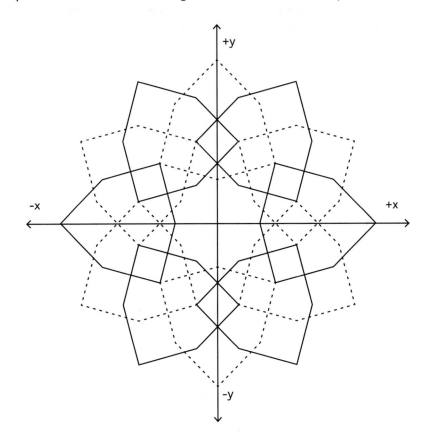

Generating the Motif

```
//scale factor
let a = 60;

function setup() {
  createCanvas(400, 400);
  angleMode(DEGREES);
  noLoop();
  noFill();
}

function draw() {

  b = a * cos(45);
  c = b * tan(30);

  let x0,y0,x1,y1,x2,y2;

  x0 = -b;
  y0 = 0;
  x1 = 0;
  y1 = b;
  x2 = -x0;
  y2 = 0;

  push();
    translate(width*0.5,height*0.5);
    rotate(90);
    for(let k=0;k<6;k++){
      push();
        rotate(k*60);
        translate(0,2*b*sin(60)-c+0.5*a*(sin(45)/sin(30)));
        rotate(60);
        for(let i=0;i<3;i++){
          push();
            rotate(120*i);
            translate(0,-c);
            beginShape();
            vertex(x0,-y0);
            vertex(x1,-y1);
            vertex(x2,-y2);
            endShape();
          pop();
        }
      pop();
    }
  pop();
}
```

Analyzing the Tessellation

Step 6: We need to calculate the dx, dy, and doff values in the placement.

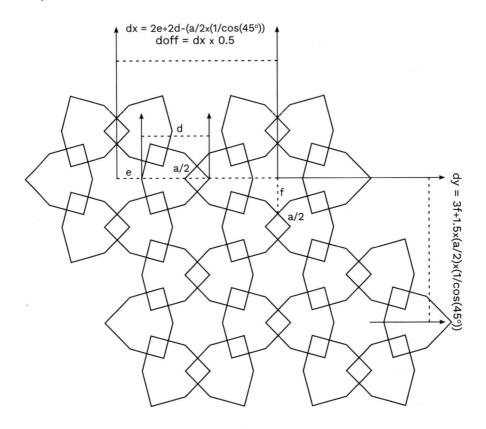

$$dx = 2e+2d-(a/2x(1/\cos(45°)))$$
$$doff = dx \times 0.5$$

d

e   a/2

f

a/2

$$dy = 3f+1.5x(a/2)x(1/\cos(45°))$$

Tessellation Code

```
//Motif class
class Motif {
    constructor(a) {
        this.a = a;
    }

    display() {
        let a = this.a;
        let b = a * cos(45);
        let c = b * tan(30);
        let d = 2 * b * sin(60) - c + 0.5 * a * (sin(45) / sin(30));

        let x0, y0, x1, y1, x2, y2;
        x0 = -b;
        y0 = 0;
        x1 = 0;
        y1 = b;
        x2 = -x0;
        y2 = 0;

        rotate(90);
        for (let k = 0; k < 6; k++) {
            push();
                rotate(k * 60);
                translate(0, d);
                rotate(60);
                for (let i = 0; i < 3; i++) {
                    push();
                        rotate(120 * i);
                        translate(0, -c);
                        beginShape();
                        vertex(x0, -y0);
                        vertex(x1, -y1);
                        vertex(x2, -y2);
                        endShape();
                    pop();
                }
            pop();
        }
    }
}
```

```
//scale factor
let a = 40;
let motif = new Motif(a);
let nRow;
let nCol;
let dx,dy;
let doff;

function setup() {
    createCanvas(800, 800);
    angleMode(DEGREES);
    noLoop();
    noFill();
    let b = a * cos(45);
    let c = b * tan(30);
    let d = b + 2 * b * sin(60);
    let e = 0.5 * a * (sin(45) / sin(30));
    let f = e * (sin(105) / sin(45));

    dx = 2 * e + 2 * d - 0.5 * a * (1 / cos(45));
    dy = 3 * f + 1.5 * (0.5 * a * (1 / cos(45)));
    doff = dx / 2;

    //approximate the nRow and nCol values
    nRow = ceil(height / dy);
    nCol = 1 + ceil(width / dx);
}

function draw() {
    push();
        for (let r = 0; r < nRow; r++) {
            for (let c = 0; c < nCol; c++) {
                push();
                    if (r % 2 == 1) {
                        //row 1,3,5,7
                        translate(-doff, 0);
                    }
                    translate(dx * c, dy * r);
                    motif.display();
                pop();
            }
        }
    pop();
}
```

Generating a Geometric Pattern Workflow #13

Observe the geometric pattern, and analyze it to distinguish its constituent repeating motif.

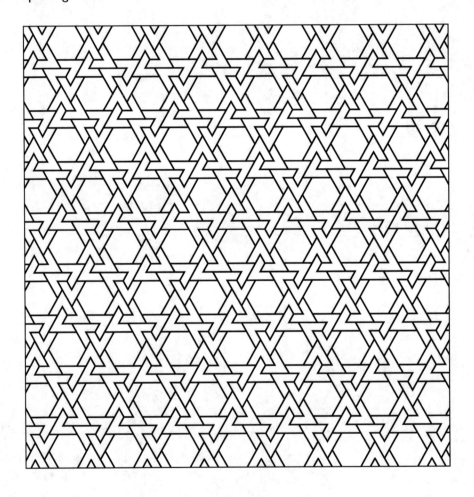

The Motif

Analyzing the Constructive Elements

Step 1: Let's find the vertex points of the constructive element.

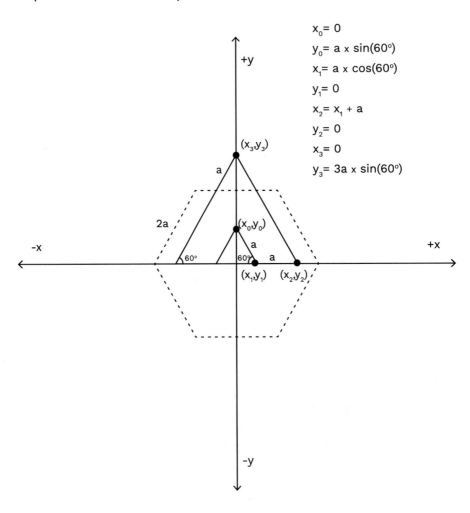

$x_0 = 0$

$y_0 = a \times \sin(60°)$

$x_1 = a \times \cos(60°)$

$y_1 = 0$

$x_2 = x_1 + a$

$y_2 = 0$

$x_3 = 0$

$y_3 = 3a \times \sin(60°)$

Step 2: The triangular shape needs to be translated to the following position.

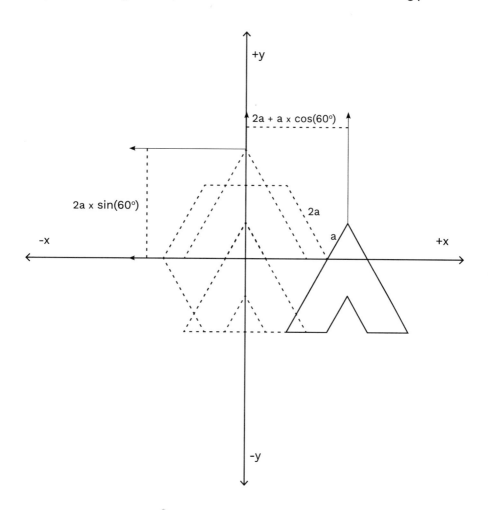

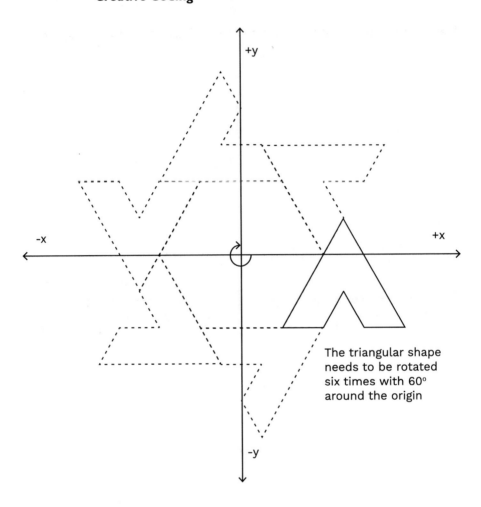

The triangular shape needs to be rotated six times with 60° around the origin

Generating the Motif

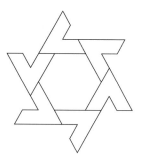

```
//scale factor
let a = 40;

function setup() {
  createCanvas(400, 400);
  angleMode(DEGREES);
  noFill();
  noLoop();
}

function draw() {

  let x0,y0,x1,y1,x2,y2,x3,y3;
  x0 = 0;
  y0 = a * sin(60);
  x1 = a * cos(60);
  y1 = 0;
  x2 = x1 + a;
  y2 = 0;
  x3 = 0;
  y3 = 3*a * sin(60);

  push();
    translate(width*0.5,height*0.5);
    for(let i=0;i<6;i++){
      push();
        rotate(i*60);
          push();
            translate(2*a+a*cos(60),2*a*sin(60));
            beginShape();
            vertex(x0,-y0);
            vertex(x1,-y1);
            vertex(x2,-y2);
            vertex(x3,-y3);
            endShape();
            //mirror on y-axis
            push();
              scale(-1,1);
              beginShape();
              vertex(x0,-y0);
              vertex(x1,-y1);
              vertex(x2,-y2);
              vertex(x3,-y3);
              endShape();
          pop();
        pop();
      pop();
    }
  pop();
```

Analyzing the Tessellation

Step 3: We need to calculate the dx, dy, and doff values in the placement.

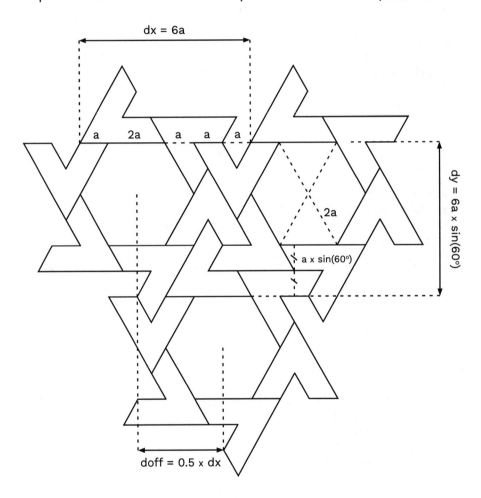

Tessellation Code

```
// Motif class
class Motif {
    constructor(a) {
        this.a = a;
    }

    display() {
        let x0, y0, x1, y1, x2, y2, x3, y3;
        x0 = 0;
        y0 = this.a * sin(60);
        x1 = this.a * cos(60);
        y1 = 0;
        x2 = x1 + this.a;
        y2 = 0;
        x3 = 0;
        y3 = 3 * this.a * sin(60);

        for (let i = 0; i < 6; i++) {
            push();
                rotate(i * 60);
                push();
                    translate(2 * this.a + this.a * cos(60), 2 * this.a * sin(60));
                    beginShape();
                    vertex(x0, -y0);
                    vertex(x1, -y1);
                    vertex(x2, -y2);
                    vertex(x3, -y3);
                    endShape();
                    //mirror on y-axis
                    push();
                        scale(-1, 1);
                        beginShape();
                        vertex(x0, -y0);
                        vertex(x1, -y1);
                        vertex(x2, -y2);
                        vertex(x3, -y3);
                        endShape();
                    pop();
                pop();
            pop();
        }
    }
}
```

```
//scale factor
let a = 16;
let motif = new Motif(a);
let nRow;
let nCol;
let dx, dy, doff;

function setup() {
    createCanvas(800, 800);
    angleMode(DEGREES);
    noFill();
    noLoop();

    dx = 6 * a;
    dy = 6 * a * sin(60);
    doff = dx * 0.5;

    //approximate the nRow and nCol values
    nCol = 1 + ceil(width / dx);
    nRow = 1 + ceil(height / dy);
}

function draw() {
    for (let c = 0; c < nCol; c++) {
        for (let r = 0; r < nRow; r++) {
            push();
                if (r % 2 == 0) {
                    //columns 0,2,4,6
                    translate(doff, 0);
                }
                translate(dx * c, dy * r);
                motif.display();
            pop();
        }
    }
}
```

Generating a Geometric Pattern Workflow #14

Observe the geometric pattern, and analyze it to distinguish its constituent repeating motif.

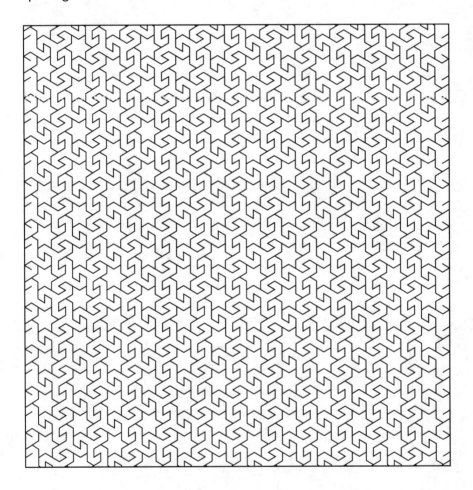

The Motif

Analyzing the Constructive Elements

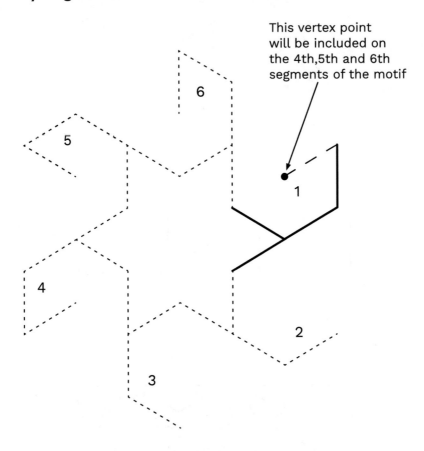

This vertex point will be included on the 4th,5th and 6th segments of the motif

Calculating Angles and Vertex Points

Step 1: Let's find the vertex points of the constructive element.

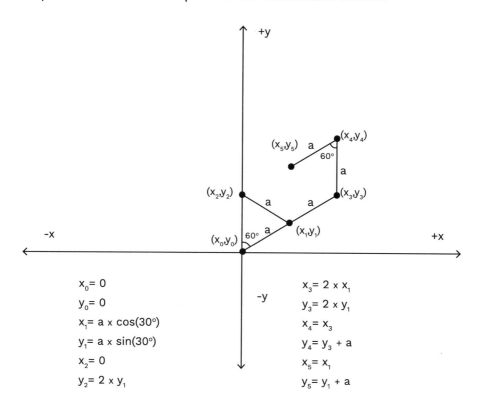

$x_0 = 0$

$y_0 = 0$

$x_1 = a \times \cos(30°)$

$y_1 = a \times \sin(30°)$

$x_2 = 0$

$y_2 = 2 \times y_1$

$x_3 = 2 \times x_1$

$y_3 = 2 \times y_1$

$x_4 = x_3$

$y_4 = y_3 + a$

$x_5 = x_1$

$y_5 = y_1 + a$

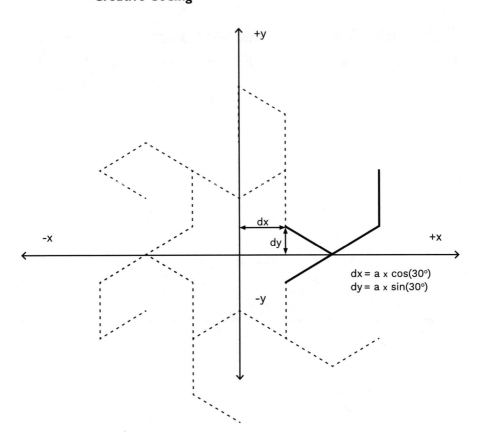

Generating the Motif

```
//scale factor
let a = 60;
function setup() {
    createCanvas(400, 400);
    angleMode(DEGREES);
    noFill();
    noLoop();
}

function draw() {
    let x0, y0, x1, y1, x2, y2, x3, y3, x4, y4, x5, y5;
    let dx = a * cos(30);
    let dy = a * sin(30);
    push();
        translate(width * 0.5, height * 0.5);
        for(let i = 0; i<6; i++){
            push();
                rotate(60 * i);
                translate(dx, dy);
                beginShape();
                x0 = 0;
                y0 = 0;
                x1 = a * cos(30);
                y1 = a * sin(30);
                x2 = 0;
                y2 = 2 * y1;
                vertex(x0, -y0);
                vertex(x1, -y1);
                vertex(x2, -y2);
                endShape();
                beginShape();
                x3 = 2 * x1;
                y3 = 2 * y1;
                x4 = x3;
                y4 = y3 + a;
                x5 = x1;
                y5 = y1 + a;
                vertex(x1, -y1);
                vertex(x3, -y3);
                vertex(x4, -y4);
                if (i >= 3) {
                    vertex(x5, -y5);
                }
                endShape();
            pop();
        }
    pop();
}
```

Analyzing the Tessellation

Step 2: We need to calculate the x-offset and y-offset values in the placement.

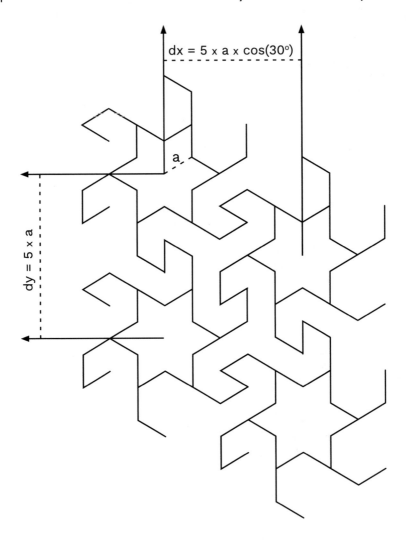

Tessellation Code

```
//Motif class
class Motif {
    constructor(a) {
        this.a = a;
    }

    display() {
        let x0, y0, x1, y1, x2, y2, x3, y3, x4, y4, x5, y5;
        let dx = this.a * cos(30);
        let dy = this.a * sin(30);
        for(let i = 0; i<6; i++){
            push();
                rotate(60 * i);
                translate(dx, dy);
                beginShape();
                x0 = 0;
                y0 = 0;
                x1 = this.a * cos(30);
                y1 = this.a * sin(30);
                x2 = 0;
                y2 = 2 * y1;
                vertex(x0, -y0);
                vertex(x1, -y1);
                vertex(x2, -y2);
                endShape();
                beginShape();
                x3 = 2 * x1;
                y3 = 2 * y1;
                x4 = x3;
                y4 = y3 + this.a;
                x5 = x1;
                y5 = y1 + this.a;
                vertex(x1, -y1);
                vertex(x3, -y3);
                vertex(x4, -y4);
                if(i >= 3) {
                    vertex(x5, -y5);
                }
                endShape();
            pop();
        }
    }
}
```

```
//scale factor
let a = 20;
let motif = new Motif(a);
let nRow;
let nCol;
let dx, dy, doff;

function setup() {
    createCanvas(800, 800);
    angleMode(DEGREES);
    noFill();
    noLoop();

    dx = 5 * a * cos(30);
    dy = 5 * a;
    doff = dy * 0.5;

  //approximate the nRow and nCol values
  nRow = 1 + ceil(height / dy);
  nCol = 1 + ceil(width / dx);
}

function draw() {
    push();
        for (let c = 0; c < nCol; c++) {
            for (let r = 0; r < nRow; r++) {
                push();
                    translate(dx * c, dy * r);
                    if (c % 2 == 1) {
                        translate(0,  doff);
                    }
                    motif.display();
                pop();
            }
        }
    pop();
}
```

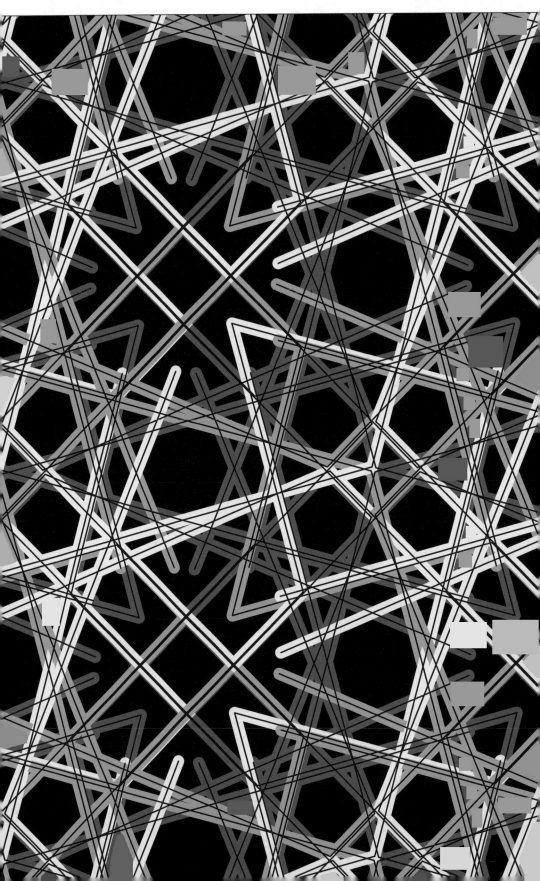

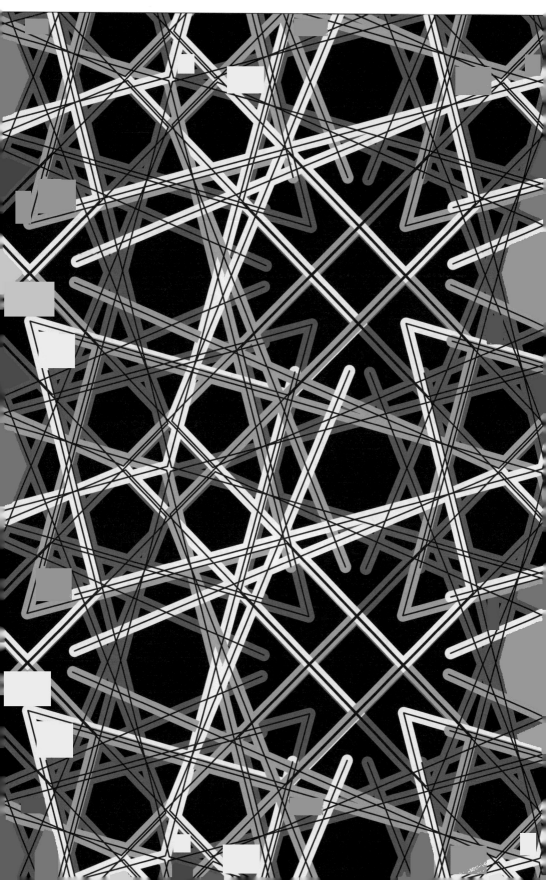

Generating a Geometric Pattern Workflow #15

Observe the geometric pattern, and analyze it to distinguish its constituent repeating motif.

The Motif

Analyzing the Constructive Elements

Observe that $\overset{\triangle}{ABC}$ is an equilateral triangle.

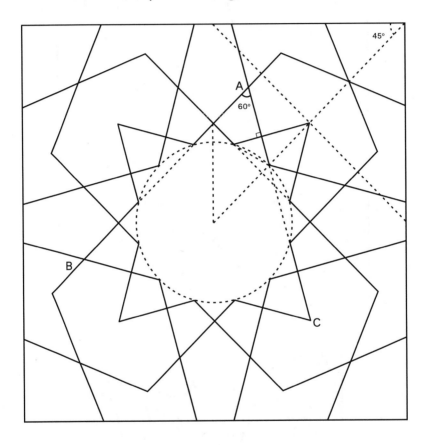

Calculating Angles and Vertex Points

Step 1: Let's find the vertex points of the constructive element.

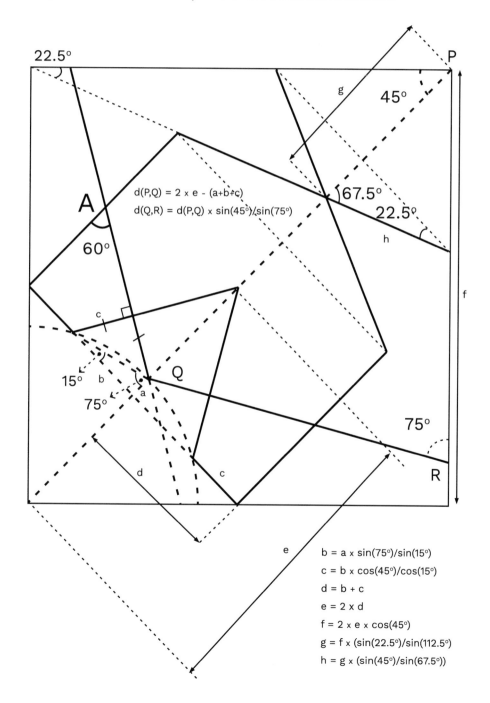

22.5°   P

g   45°

d(P,Q) = 2 x e - (a+b+c)
d(Q,R) = d(P,Q) x sin(45°)/sin(75°)

A   67.5°

60°   22.5°

h

c

15° b

75°   Q

a   75°

d   c   R

e   b = a x sin(75°)/sin(15°)
c = b x cos(45°)/cos(15°)
d = b + c
e = 2 x d
f = 2 x e x cos(45°)
g = f x (sin(22.5°)/sin(112.5°))
h = g x (sin(45°)/sin(67.5°))

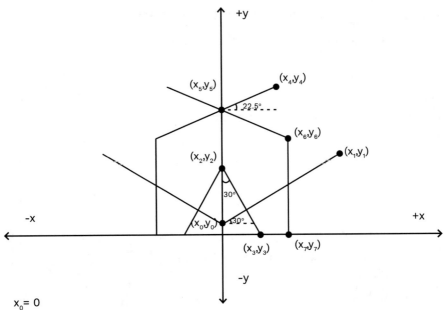

$x_0 = 0$

$y_0 = a$

$x_1 = (2 \times e - (a+b+c) \times (\sin(45°)/\sin(75°))) \times \cos(30°)$

$y_1 = (2 \times e - (a+b+c) \times (\sin(45°)/\sin(75°))) \times \sin(30°) + a$

$x_2 = 0$

$y_2 = d$

$x_3 = \tan(30°) \times y_2$

$y_3 = 0$

$x_4 = b + c$

$y_4 = 0$

$x_5 = x_4$

$y_5 = (e - g)/2 + y_2$

$x_6 = 0$

$y_6 = 3 \times d - g$

$x_7 = h \times \cos(22.5°)$

$y_7 = h \times \sin(22.5°) + 3 \times d - g$

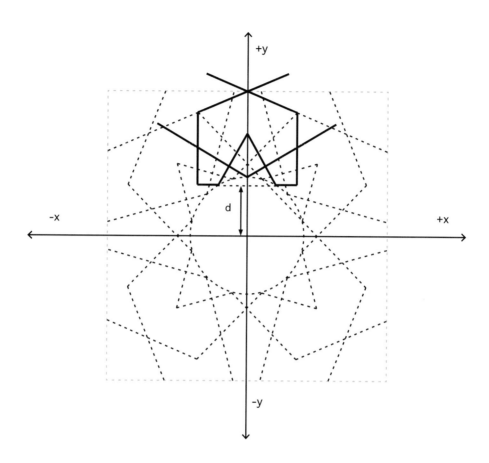

Generating the Motif

```
//scale factor
let a = 10;
let b,c,d,e,f,g,h;

function setup() {
  createCanvas(400, 400);
  angleMode(DEGREES);
  noFill();
  noLoop();
  a = 10;
  b = a * sin(75)/sin(15);
  c = b * cos(45)/cos(15);
  d = b + c;
  e = 2 * d;
  f = (2 * e) * cos(45);
  g = f * (sin(22.5)/sin(112,5));
  h = g * (sin(45)/sin(67,5));
}
```

```
function draw() {
  let x0,y0,x1,y1,x2,y2,x3,y3,x4,y4,x5,y5,x6,y6,x7,y7;
  x0 = 0;
  y0 = a;
  x1 = ((2.0*e-(a+b+c)) * (sin(45)/sin(75))) * cos(30);
  y1 = ((2.0*e-(a+b+c)) * (sin(45)/sin(75))) * sin(30) + a;
  x2 = 0;
  y2 = d;
  x3 = tan(30) * y2;
  y3 = 0;
  x4 = b + c;
  y4 = 0;
  x5 = x4;
  y5 = (e - g)/2 + y2;
  x6 = 0;
  y6 = 3 * d - g;
  x7 = h * cos(22.5);
  y7 = h * sin(22.5) + 3 * d - g;
```

```
push();
  translate(width*0.5,height*0.5);
  rotate(45);
  for(let j=0;j<4;j++){
    push();
      rotate(j*90);
        push();
          translate(0,-d);
          let mirror = 1;
          for(let i=0;i<2;i++){
            push();
              scale(mirror,1);
              beginShape();
              vertex(x0,-y0);
              vertex(x1,-y1);
              endShape();
              beginShape();
              vertex(x2,-y2);
              vertex(x3,-y3);
              vertex(x4,-y4);
              vertex(x5,-y5);
              vertex(x6,-y6);
              vertex(x7,-y7);
              endShape();
            pop();
            mirror = mirror * -1;
          }
        pop();
      pop();
    }
  pop();
}
```

Analyzing the Tessellation

Step 2: We need to calculate the dx and dy values in the placement.

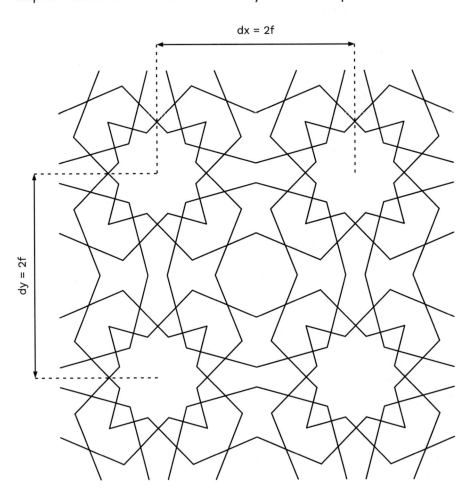

Tessellation Code

```
//Motif class
class Motif {
    constructor(a) {
        this.a = a;
    }

    display() {
        let a, b, c, d, e, f, g, h;
        a = this.a;
        b = (a * sin(75)) / sin(15);
        c = (b * cos(45)) / cos(15);
        d = b + c;
        e = 2 * d;
        f = (2 * e) * cos(45);
        g = f * (sin(22.5) / sin(112, 5));
        h = g * (sin(45) / sin(67, 5));

        let x0, y0, x1, y1, x2, y2, x3, y3, x4, y4, x5, y5, x6, y6, x7, y7;
        x0 = 0;
        y0 = a;
        x1 = (2.0 * e - (a + b + c)) * (sin(45) / sin(75)) * cos(30);
        y1 = (2.0 * e - (a + b + c)) * (sin(45) / sin(75)) * sin(30) + a;
        x2 = 0;
        y2 = d;
        x3 = tan(30) * y2;
        y3 = 0;
        x4 = b + c;
        y4 = 0;
        x5 = x4;
        y5 = (e - g) / 2 + y2;
        x6 = 0;
        y6 = 3 * d - g;
        x7 = h * cos(22.5);
        y7 = h * sin(22.5) + 3 * d - g;

        push();
            rotate(45);
            for (let j = 0; j < 4; j++) {
                push();
                    rotate(j * 90);
                    push();
                        translate(0, -d);
                        let mirror = 1;
                        for (let i = 0; i < 2; i++){
                            push();
```

```
                                    scale(mirror,1);
                                    beginShape();
                                    vertex(x0, -y0);
                                    vertex(x1, -y1);
                                    endShape();
                                    beginShape();
                                    vertex(x2, -y2);
                                    vertex(x3, -y3);
                                    vertex(x4, -y4);
                                    vertex(x5, -y5);
                                    vertex(x6, -y6);
                                    vertex(x7, -y7);
                                    endShape();
                                pop();
                                mirror = mirror * -1;
                            }
                        pop();
                    pop();
                }
            pop();
        }
}

//scale factor
let a = 5;
let motif = new Motif(a);
let nRow;
let nCol;
let xOff, yOff;

function setup() {
    createCanvas(800, 800);
    angleMode(DEGREES);
    noFill();
    noLoop();

    let b, c, d, e, f, g, h;
    b = (a * sin(75)) / sin(15);
    c = (b * cos(45)) / cos(15);
    d = b + c;
    e = 2 * d;
    f = (2 * e) * cos(45);

    xOff = 2*f;
    yOff = 2*f;
```

```
    //approximate the nRow and nCol values
    nRow = ceil(height / xOff);
    nCol = ceil(width / yOff);
}

function draw() {
    push();
        for (let c = 0; c < nCol; c++) {
            for (let r = 0; r < nRow; r++) {
                push();
                    translate(xOff * c, yOff * r);
                    motif.display();
                pop();
            }
        }
    pop();
}
```

Generating a Geometric Pattern Workflow #16

Observe the geometric pattern, and analyze it to distinguish its constituent repeating motif.

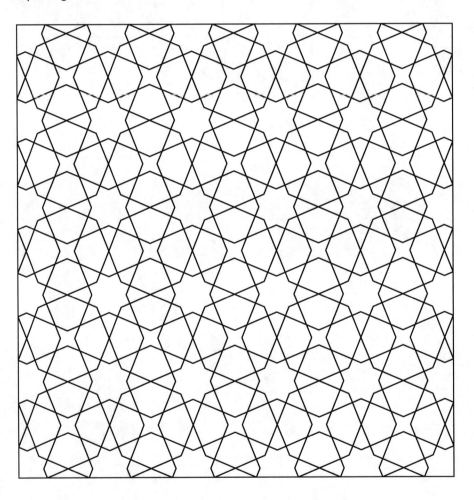

The Motif

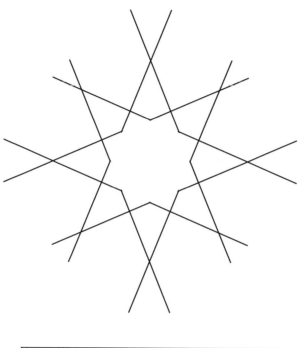

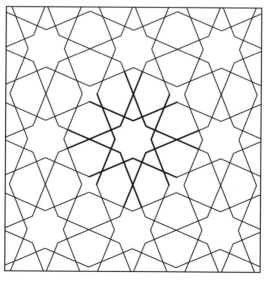

Analyzing the Constructive Elements

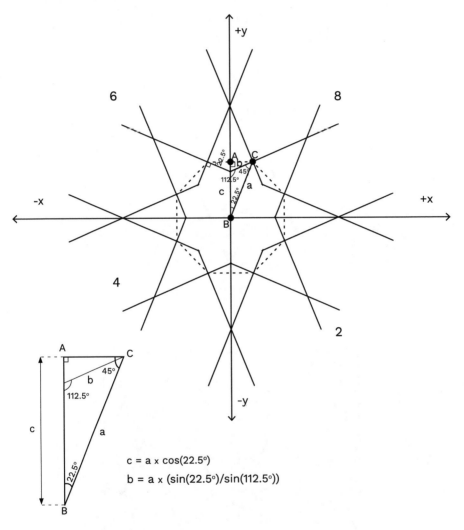

$$c = a \times \cos(22.5°)$$
$$b = a \times (\sin(22.5°)/\sin(112.5°))$$

See that segments 2, 4 , 6, and 8 are shorter to generate a star shape in between the motif tessellations.

Let's focus on the segments closely.

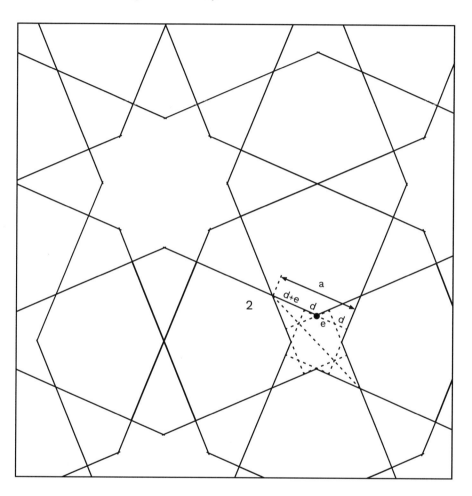

a = 3d + 2e

d = e x cos(45°)

we infer that;

e = a / (3 x cos(45°)+2)

Calculating Angles and Vertex Points

Step 1: Let's find the vertex points of the constructive element.

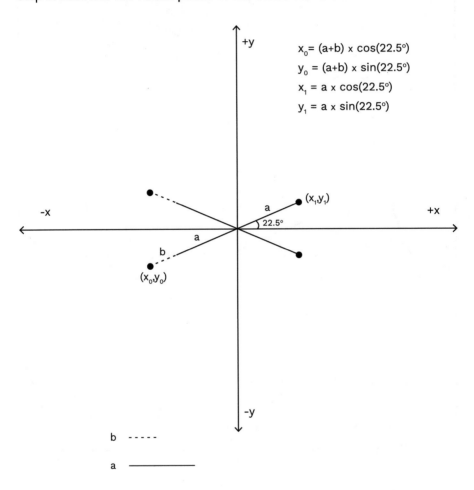

$$x_0 = (a+b) \times \cos(22.5°)$$
$$y_0 = (a+b) \times \sin(22.5°)$$
$$x_1 = a \times \cos(22.5°)$$
$$y_1 = a \times \sin(22.5°)$$

Step 2: Special condition for the segments 2, 4, 6, and 8.

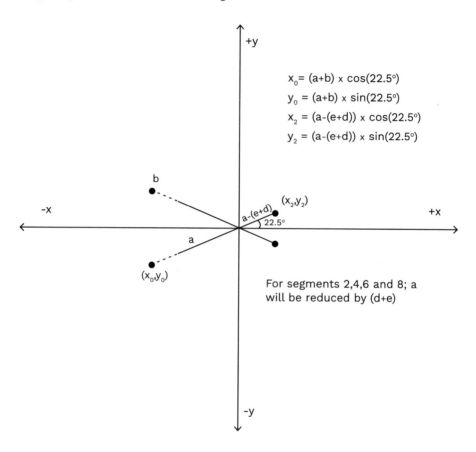

$$x_0 = (a+b) \times \cos(22.5°)$$
$$y_0 = (a+b) \times \sin(22.5°)$$
$$x_2 = (a-(e+d)) \times \cos(22.5°)$$
$$y_2 = (a-(e+d)) \times \sin(22.5°)$$

For segments 2,4,6 and 8; a will be reduced by (d+e)

Generating the Motif

```
//scale factor
let a = 60;
let b, c, d, e;

function setup() {
    createCanvas(400, 400);
    angleMode(DEGREES);
    noLoop();
    noFill();
    c = a * cos(22.5);
    b = a * (sin(22.5) / sin(112.5));
    e = a / (3 * cos(45) + 2);
    d = e * cos(45);
}
function draw() {
    let x0, y0, x1, y1, x2, y2;
    x0 = (a + b) * cos(22.5);
    y0 = (a + b) * sin(22.5);
    x1 = a * cos(22.5);
    y1 = a * sin(22.5);
    x2 = (a - (e + d)) * cos(22.5);
    y2 = (a - (e + d)) * sin(22.5);
    push();
        translate(width * 0.5, height * 0.5);
        for (let i = 0; i < 8; i++) {
            push();
                rotate(i * 45);
                translate(2 * c, 0);
                let mirror = 1;
                for (let m = 0; m < 2; m++) {
                    push();
                        //reflect on x axis
                        scale(1,mirror);
                        beginShape();
                        vertex(-x0, -y0);
                        //segments2,4,6,8
                        if (i % 2 == 1) {
                            vertex(x2, y2);
                        }else{
                            vertex(x1, y1);
                        }
                        endShape();
                        mirror = -1;
                    pop();
                }
            pop();
        }
    pop();
}
```

Analyzing the Tessellation

Step 3: We need to calculate the dx and dy values in the placement.

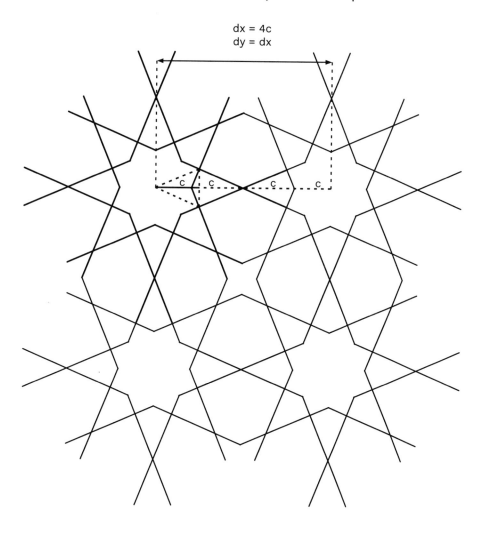

Tessellation Code

```
// Motif class
class Motif {
  constructor(a) {
    this.a = a;
  }

  display() {
    let c = this.a * cos(22.5);
    let b = this.a * (sin(22.5)/sin(112.5));
    let e = this.a / (3*cos(45)+2);
    let d = e * cos(45);

    let x0,y0,x1,y1,x2,y2;
    x0 = (a+b)*cos(22.5);
    y0 = (a+b)*sin(22.5);
    x1 = a*cos(22.5);
    y1 = a*sin(22.5);
    x2 = (a-(e+d)) * cos(22.5);
    y2 = (a-(e+d)) * sin(22.5);

    for(let i=0;i<8;i++){
      push();
        rotate(i*45);
        translate(2*c,0);
        let mirror = 1;
        for(let m=0;m<2;m++){
          push();
            //reflect on x axis
            scale(1,mirror);
            beginShape();
            vertex(-x0,-y0);
            //segments 2,4,6,8
            if(i%2==1){
              vertex(x2,y2);
            }else{
              vertex(x1,y1);
            }
            endShape();
            mirror = -1;
          pop();
        }
      pop();
    }
  }
}
```

```
//scale factor
let a = 40;
let motif = new Motif(a);
let nRow;
let nCol;
let dx, dy;

function setup() {
  createCanvas(800, 800);
  angleMode(DEGREES);
  noLoop();
  noFill();

  let c =  a * cos(22.5);

  dx = 4*c;
  dy = dx;

  //approximate the nRow and nCol values
  nRow = ceil(height / dy);
  nCol = ceil(width / dx);
}

function draw() {
  for (let c = 0; c < nCol; c++) {
    for (let r = 0; r < nRow; r++) {
      push();
        translate(dx * c, dy * r);
        motif.display();
      pop();
    }
  }
}
```

Generating a Geometric Pattern Workflow #17

Observe the geometric pattern, and analyze it to distinguish its constituent repeating motif.

The Motif

Analyzing the Constructive Elements

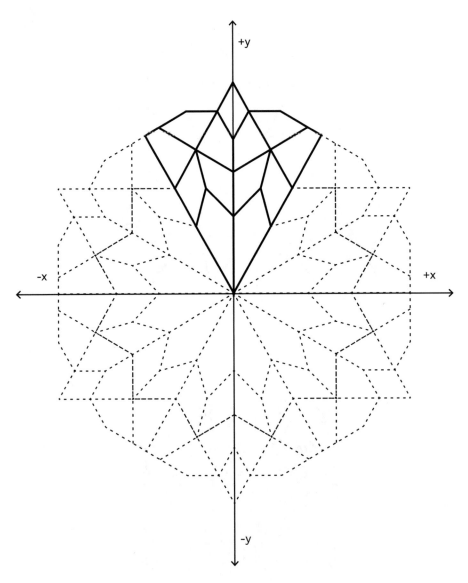

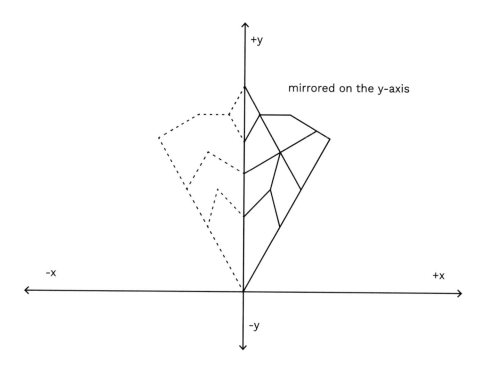

mirrored on the y-axis

Step 1: Let's focus on the first shape.

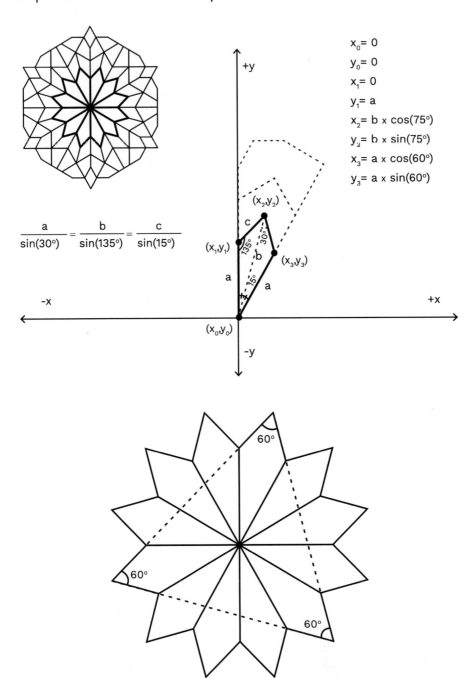

$x_0 = 0$

$y_0 = 0$

$x_1 = 0$

$y_1 = a$

$x_2 = b \times \cos(75°)$

$y_2 = b \times \sin(75°)$

$x_3 = a \times \cos(60°)$

$y_3 = a \times \sin(60°)$

$$\frac{a}{\sin(30°)} = \frac{b}{\sin(135°)} = \frac{c}{\sin(15°)}$$

Generating the Motif: Shape 1

```
//scale factor
let a = 60;
let b;

function setup() {
  createCanvas(400, 400);
  angleMode(DEGREES);
  noFill();
  noLoop();
  b = a * (sin(135) / sin(30));
}

function draw() {
  let x0,y0,x1,y1,x2,y2,x3,y3;
  x0 = 0;
  y0 = 0;
  x1 = 0;
  y1 = a;
  x2 = b * cos(75);
  y2 = b * sin(75);
  x3 = a * cos(60);
  y3 = a * sin(60);

  push();
    translate(width*0.5, height*0.5);
    //Shape 1
    for (let i = 0; i < 12; i++) {
      push();
        rotate(30 * i);
        beginShape();
        vertex(x0, y0);
        vertex(x1, y1);
        vertex(x2, y2);
        vertex(x3, y3);
        endShape(CLOSE);
      pop();
    }
  pop();
}
```

Analyzing the Constructive Elements

Step 2: Let's focus on the second shape.

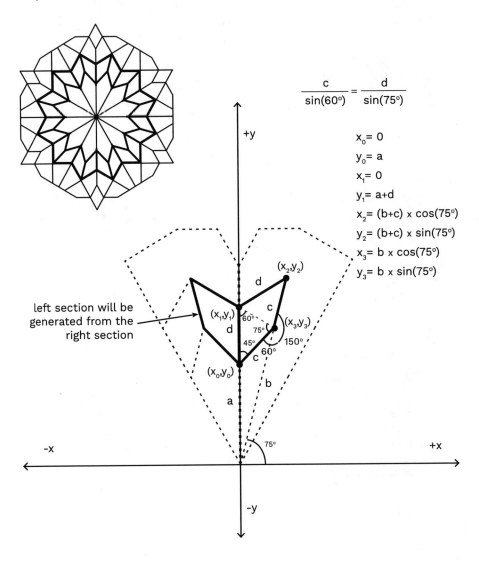

$$\frac{c}{\sin(60°)} = \frac{d}{\sin(75°)}$$

$x_0 = 0$

$y_0 = a$

$x_1 = 0$

$y_1 = a+d$

$x_2 = (b+c) \times \cos(75°)$

$y_2 = (b+c) \times \sin(75°)$

$x_3 = b \times \cos(75°)$

$y_3 = b \times \sin(75°)$

left section will be generated from the right section

Generating the Motif: Shape 2 Added

```
//scale factor
let a = 60;
let b,c,d;

function setup() {
  createCanvas(400, 400);
  angleMode(DEGREES);
  noFill();
  noLoop();
  b = a * (sin(135) / sin(30));
  c = a * (sin(15) / sin(30));
  d = c * (sin(75) / sin(60));
}

function draw() {
  let x0,y0,x1,y1,x2,y2,x3,y3;
  x0 = 0;
  y0 = 0;
  x1 = 0;
  y1 = a;
  x2 = b * cos(75);
  y2 = b * sin(75);
  x3 = a * cos(60);
  y3 = a * sin(60);

  push();
    translate(width*0.5, height*0.5);
    //Shape 1
    for (let i = 0; i < 12; i++) {
      push();
        rotate(30 * i);
        beginShape();
        vertex(x0, -y0);
        vertex(x1, -y1);
        vertex(x2, -y2);
        vertex(x3, -y3);
        endShape(CLOSE);
      pop();
    }
```

Generating the Motif: Shape 2 Added

```
x0 = 0;
y0 = a;
x1 = 0;
y1 = a+d;
x2 = (b+c) * cos(75);
y2 = (b+c) * sin(75);
x3 = b * cos(75);
y3 = b * sin(75);

//Shape 2
for (let i = 0; i < 12; i++) {
  //right section
  push();
    rotate(30 * i);
    beginShape();
    vertex(x0, -y0);
    vertex(x1, -y1);
    vertex(x2, -y2);
    vertex(x3, -y3);
    endShape(CLOSE);
    //left section
    scale(-1,1);
    beginShape();
    vertex(x0, -y0);
    vertex(x1, -y1);
    vertex(x2, -y2);
    vertex(x3, -y3);
    endShape(CLOSE);
  pop();
}
pop();
}
```

Analyzing the Constructive Elements

Step 3: Let's focus on the third shape.

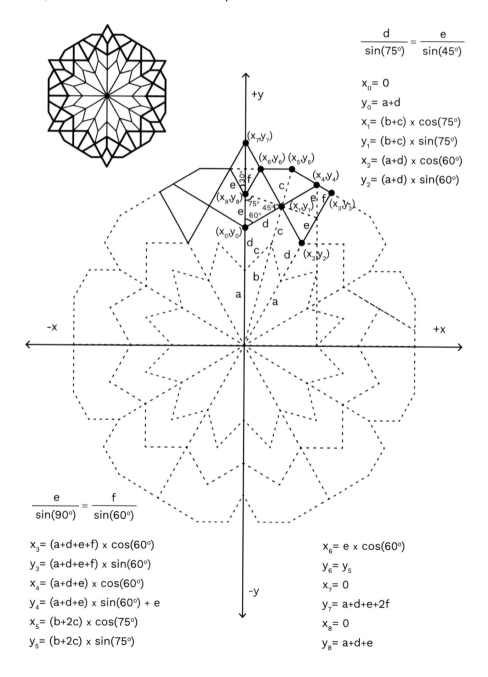

$$\frac{d}{\sin(75°)} = \frac{e}{\sin(45°)}$$

$x_0 = 0$

$y_0 = a+d$

$x_1 = (b+c) \times \cos(75°)$

$y_1 = (b+c) \times \sin(75°)$

$x_2 = (a+d) \times \cos(60°)$

$y_2 = (a+d) \times \sin(60°)$

$$\frac{e}{\sin(90°)} = \frac{f}{\sin(60°)}$$

$x_3 = (a+d+e+f) \times \cos(60°)$

$y_3 = (a+d+e+f) \times \sin(60°)$

$x_4 = (a+d+e) \times \cos(60°)$

$y_4 = (a+d+e) \times \sin(60°) + e$

$x_5 = (b+2c) \times \cos(75°)$

$y_5 = (b+2c) \times \sin(75°)$

$x_6 = e \times \cos(60°)$

$y_6 = y_5$

$x_7 = 0$

$y_7 = a+d+e+2f$

$x_8 = 0$

$y_8 = a+d+e$

Generating the Motif: Shape 3 Added

```
//scale factor
let a = 60;
let b,c,d,e,f;

function setup() {
  createCanvas(400, 400);
  angleMode(DEGREES);
  noFill();
  noLoop();
  b = a * (sin(135) / sin(30));
  c = a * (sin(15) / sin(30));
  d = c * (sin(75) / sin(60));
  e = d * (sin(45) / sin(75));
  f = e * (sin(60) / sin(90));
}
function draw() {
  let x0,y0,x1,y1,x2,y2,x3,y3,x4,
  y4,x5,y5,x6,y6,x7,y7,x8,y8;
  //Shape 1 vertices
  x0 = 0;
  y0 = 0;
  x1 = 0;
  y1 = a;
  x2 = b * cos(75);
  y2 = b * sin(75);
  x3 = a * cos(60);
  y3 = a * sin(60);
  x4 = a * cos(60);
  y4 = a * sin(60);
  push();
    translate(width*0.5, height*0.5);
    //Shape 1
    for (let i = 0; i < 12; i++) {
      push();
        rotate(30 * i);
        beginShape();
        vertex(x0, -y0);
        vertex(x1, -y1);
        vertex(x2, -y2);
        vertex(x3, -y3);
        endShape(CLOSE);
      pop();
    }
}
```

Generating the Motif: Shape 3 Added

```
//Shape 2 vertices
x0 = 0;
y0 = a;
x1 = 0;
y1 = a+d;
x2 = (b+c) * cos(75);
y2 = (b+c) * sin(75);
x3 = b * cos(75);
y3 = b * sin(75);

//Shape 2
for (let i = 0; i < 12; i++) {
  //right section
  push();
    rotate(30 * i);
    beginShape();
    vertex(x0, -y0);
    vertex(x1, -y1);
    vertex(x2, -y2);
    vertex(x3, -y3);
    endShape(CLOSE);
    //left section
    scale(-1,1);
    beginShape();
    vertex(x0, -y0);
    vertex(x1, -y1);
    vertex(x2, -y2);
    vertex(x3, -y3);
    endShape(CLOSE);
  pop();
}

//Shape 3 vertices
x0 = 0;
y0 = a+d;
x1 = (b+c) * cos(75);
y1 = (b+c) * sin(75);
x2 = (a+d) * cos(60);
y2 = (a+d) * sin(60);
x3 = (a+d+e+f) * cos(60);
y3 = (a+d+e+f) * sin(60);
x4 = (a+d+e) * cos(60);
y4 = (a+d+e) * sin(60) + e;
x5 = (b+2*c) * cos(75);
y5 = (b+2*c) * sin(75);;
x6 = e * cos(60);
y6 = y5;
```

Generating the Motif: Shape 3 Added

```
x7 = 0;
y7 = a+d+e+2*f;
x8 = 0;
y8 = a+d+e;

 //Shape 3
 for (let i = 0; i < 6; i++) {
  //right section
  push();
    rotate(60 * i);
    beginShape();
    vertex(x0, -y0);
    vertex(x1, -y1);
    vertex(x2, -y2);
    vertex(x3, -y3);
    vertex(x5, -y5);
    vertex(x6, -y6);
    vertex(x7, -y7);
    endShape();
    beginShape();
    vertex(x0, -y0);
    vertex(x4, -y4);
    endShape();
    beginShape();
    vertex(x2, -y2);
    vertex(x6, -y6);
    vertex(x8, -y8);
    vertex(x0, -y0);
    endShape();
    //left section
    scale(-1,1);
    beginShape();
    vertex(x0, -y0);
    vertex(x1, -y1);
    vertex(x2, -y2);
    vertex(x3, -y3);
    vertex(x5, -y5);
    vertex(x6, -y6);
    vertex(x7, -y7);
    endShape();
    beginShape();
    vertex(x0, -y0);
    vertex(x4, -y4);
    endShape();
    beginShape();
```

```
      vertex(x2, -y2);
      vertex(x6, -y6);
      vertex(x8, -y8);
      vertex(x0, -y0);
      endShape();
    pop();
  }
  pop();
}
```

Analyzing the Tessellation

Step 4: We need to calculate the dx, dy, and doff values in the placement.

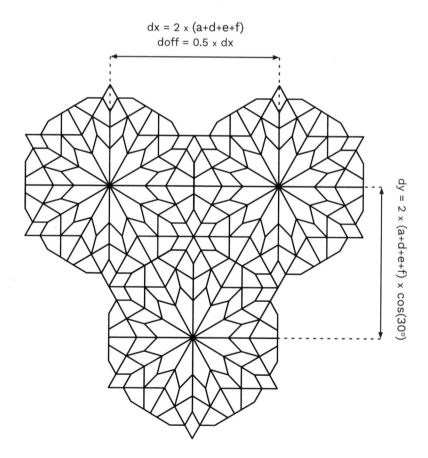

Tessellation Code

```
// Motif class
class Motif {
    constructor(a) {
        this.a = a;
    }

    display() {
        let b = this.a * (sin(135) / sin(30));
        let c = this.a * (sin(15) / sin(30));
        let d = c * (sin(75) / sin(60));
        let e = d * (sin(45) / sin(75));
        let f = e * (sin(60) / sin(90));

        let x0,y0,x1,y1,x2,y2,x3,y3,x4,y4,x5,y5,x6,y6,x7,y7,x8,y8;

        //Shape 1 vertices
        x0 = 0;
        y0 = 0;
        x1 = 0;
        y1 = this.a;
        x2 = b * cos(75);
        y2 = b * sin(75);
        x3 = this.a * cos(60);
        y3 = this.a * sin(60);
        x4 = this.a * cos(60);
        y4 = this.a * sin(60);

        //Shape 1
        for (let i = 0; i < 12; i++) {
            push();
                rotate(30 * i);
                beginShape();
                vertex(x0, -y0);
                vertex(x1, -y1);
                vertex(x2, -y2);
                vertex(x3, -y3);
                endShape(CLOSE);
            pop();
        }

        //Shape 2 vertices
        x0 = 0;
        y0 = this.a;
        x1 = 0;
        y1 = this.a + d;
        x2 = (b + c) * cos(75);
```

Tessellation Code

```
y2 = (b + c) * sin(75);
x3 = b * cos(75);
y3 = b * sin(75);

//Shape 2
for (let i = 0; i < 12; i++) {
    //right section
    push();
        rotate(30 * i);
        beginShape();
        vertex(x0, -y0);
        vertex(x1, -y1);
        vertex(x2, -y2);
        vertex(x3, -y3);
        endShape(CLOSE);
        //left section
        scale(-1, 1);
        beginShape();
        vertex(x0, -y0);
        vertex(x1, -y1);
        vertex(x2, -y2);
        vertex(x3, -y3);
        endShape(CLOSE);
    pop();
}

//Shape 3 vertices
x0 = 0;
y0 = this.a + d;
x1 = (b + c) * cos(75);
y1 = (b + c) * sin(75);
x2 = (this.a + d) * cos(60);
y2 = (this.a + d) * sin(60);
x3 = (this.a + d + e + f) * cos(60);
y3 = (this.a + d + e + f) * sin(60);
x4 = (this.a + d + e) * cos(60);
y4 = (this.a + d + e) * sin(60) + e;
x5 = (b + 2 * c) * cos(75);
y5 = (b + 2 * c) * sin(75);
x6 = e * cos(60);
y6 = y5;
x7 = 0;
y7 = this.a + d + e + 2 * f;
x8 = 0;
y8 = this.a + d + e;
```

Tessellation Code

```
//Shape 3
for (let i = 0; i < 6; i++) {
    //right section
    push();
        rotate(60 * i);
        beginShape();
        vertex(x0, -y0);
        vertex(x1, -y1);
        vertex(x2, -y2);
        vertex(x3, -y3);
        vertex(x5, -y5);
        vertex(x6, -y6);
        vertex(x7, -y7);
        endShape();
        beginShape();
        vertex(x0, -y0);
        vertex(x4, -y4);
        endShape();
        beginShape();
        vertex(x2, -y2);
        vertex(x6, -y6);
        vertex(x8, -y8);
        vertex(x0, -y0);
        endShape();
        //left section
        scale(-1, 1);
        beginShape();
        vertex(x0, -y0);
        vertex(x1, -y1);
        vertex(x2, -y2);
        vertex(x3, -y3);
        vertex(x5, -y5);
        vertex(x6, -y6);
        vertex(x7, -y7);
        endShape();
        beginShape();
        vertex(x0, -y0);
        vertex(x4, -y4);
        endShape();
        beginShape();
        vertex(x2, -y2);
        vertex(x6, -y6);
        vertex(x8, -y8);
        vertex(x0, -y0);
        endShape();
```

Tessellation Code

```
            pop();
        }
      }
}

//scale factor
let a = 40;
let motif = new Motif(a);
let nRow;
let nCol;
let dx, dy, doff;

function setup() {
    createCanvas(800, 800);
    angleMode(DEGREES);
    noFill();
    noLoop();
    let b = a * (sin(135) / sin(30));
    let c = a * (sin(15) / sin(30));
    let d = c * (sin(75) / sin(60));
    let e = d * (sin(45) / sin(75));
    let f = e * (sin(60) / sin(90));

    dx = 2 * (a + d + e + f);
    dy = 2 * (a + d + e + f) * cos(30);
    doff = a + d + e + f;

    //approximate the nRow and nCol values
    nRow = 1 + ceil(height / dy);
    nCol = ceil(width / dx);
}

function draw() {
    for (let c = 0; c < nCol; c++) {
        for (let r = 0; r < nRow; r++) {
            push();
                translate(dx * c, dy * r);
                if (r % 2 == 1) {
                    translate(doff, 0);
                }
                motif.display();
            pop();
        }
    }
}
```

Generating a Geometric Pattern Workflow #18

Observe the geometric pattern, and analyze it to distinguish its constituent repeating motif.

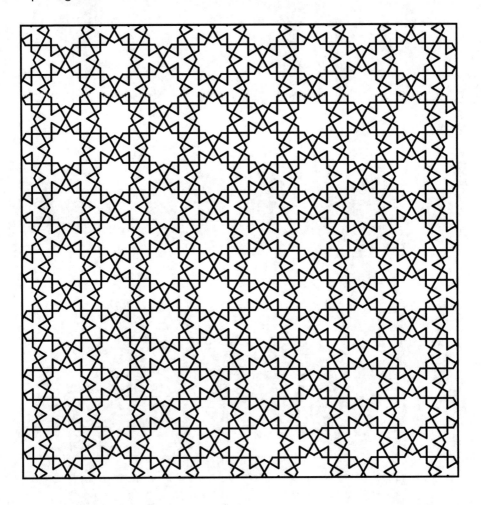

The Motif

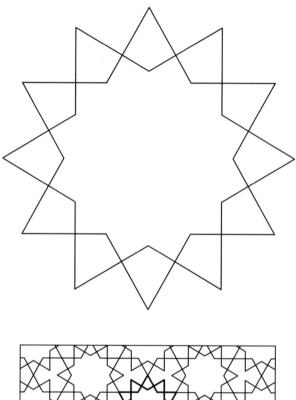

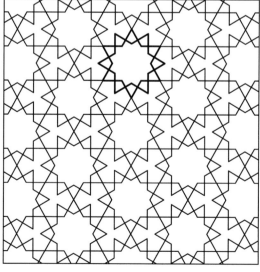

Analyzing the Constructive Elements

There are two types of repeating shapes.

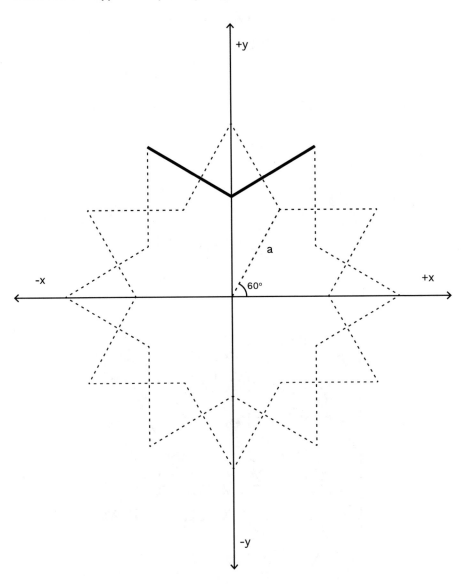

Step 1: Let's find the vertex points of the constructive element.

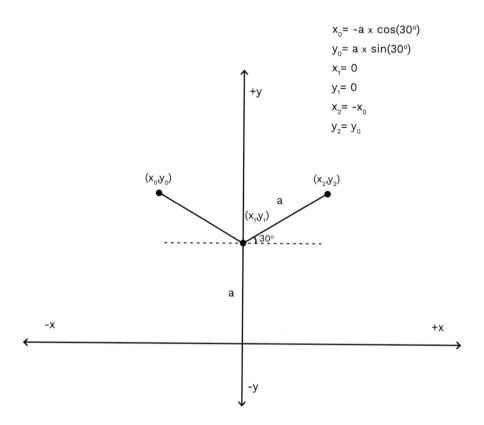

$$x_0 = -a \times \cos(30°)$$
$$y_0 = a \times \sin(30°)$$
$$x_1 = 0$$
$$y_1 = 0$$
$$x_2 = -x_0$$
$$y_2 = y_0$$

Generating the Motif

```
//scale factor
let a = 80;

function setup() {
  createCanvas(400, 400);
  angleMode(DEGREES);
  noFill();
  noLoop();
}

function draw() {

  let x0, y0, x1, y1;
  x0 = 0;
  y0 = 0;
  x1 = a * cos(30);
  y1 = a * sin(30);

  push();
    translate(width * 0.5, height * 0.5);
    for (let i = 0; i < 12; i++) {
      push();
        rotate(i*30);
        translate(0, -a);
        beginShape();
        vertex(x1, -y1);
        vertex(x0, y0);
        vertex(-x1, -y1);
        endShape();
      pop();
    }
  pop();
}
```

Analyzing the Tessellation

Step 2: We need to calculate the dx and dy values in the placement.

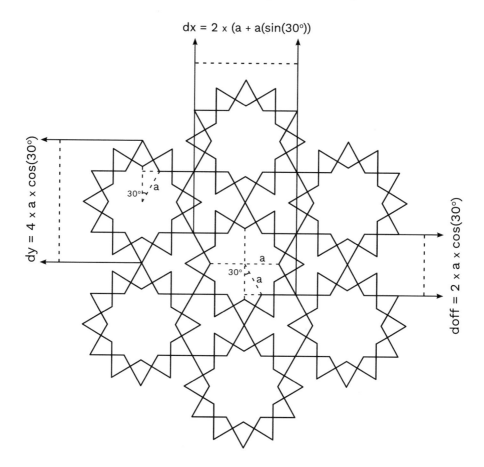

Tessellation Code

```
// Motif class
class Motif {
  constructor(a) {
    this.a = a;
  }
  display() {

    let x0, y0, x1, y1;
    x0 = 0;
    y0 = 0;
    x1 = a * cos(30);
    y1 = a * sin(30);

    for (let i = 0; i < 12; i++) {
      push();
        rotate(i*30);
        translate(0, -a);
        beginShape();
        vertex(x1, -y1);
        vertex(x0, -y0);
        vertex(-x1, -y1);
        endShape();
      pop();
      }
  }
}

//scale factor
let a = 32;
let motif = new Motif(a);
let nRow;
let nCol;
let dx, dy, doff;

function setup() {
  createCanvas(800, 800);
  angleMode(DEGREES);
  noFill();
  noLoop();

  dx = 2*(a+a*sin(30));
  dy = 4*a*cos(30);
  doff = 2*a*cos(30);
```

```
  //approximate the nRow and nCol values
  nRow = 1+ceil(height / dy);
  nCol = 1+ceil(width / dx);

}
function draw() {

  for (let c = 0; c < nCol; c++) {
    for (let r = 0; r < nRow; r++) {
      push();
        if (c % 2 == 1) {
          //columns 1,3,5,7
          translate(0, doff);
        }
        translate(dx * c, dy * r);
        motif.display();
      pop();
    }
  }
}
```

Generating a Geometric Pattern Workflow #19

Observe the geometric pattern, and analyze it to distinguish its constituent repeating motif.

The Motif

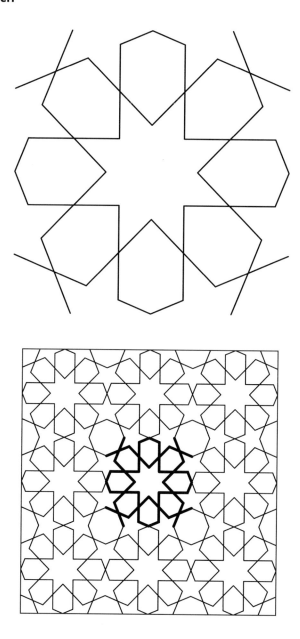

Analyzing the Constructive Elements

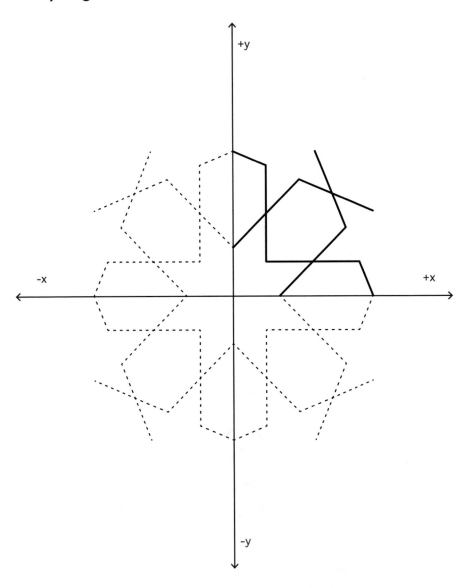

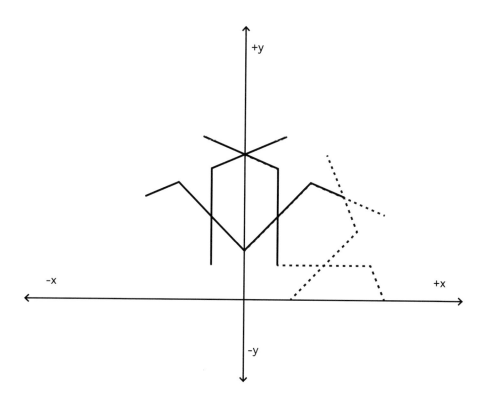

Step 1: Let's find the vertex points of the constructive element.

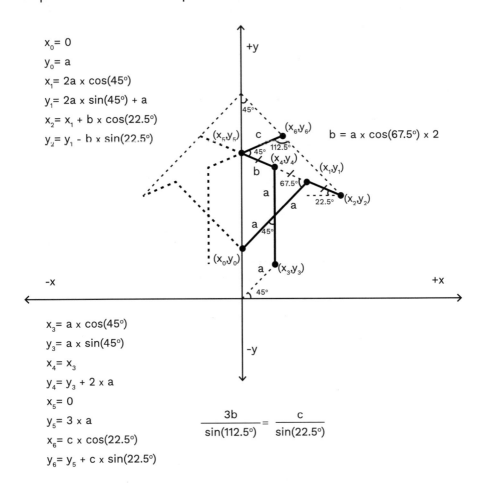

$x_0 = 0$

$y_0 = a$

$x_1 = 2a \times \cos(45°)$

$y_1 = 2a \times \sin(45°) + a$

$x_2 = x_1 + b \times \cos(22.5°)$

$y_2 = y_1 - b \times \sin(22.5°)$

$b = a \times \cos(67.5°) \times 2$

$x_3 = a \times \cos(45°)$

$y_3 = a \times \sin(45°)$

$x_4 = x_3$

$y_4 = y_3 + 2 \times a$

$x_5 = 0$

$y_5 = 3 \times a$

$x_6 = c \times \cos(22.5°)$

$y_6 = y_5 + c \times \sin(22.5°)$

$$\frac{3b}{\sin(112.5°)} = \frac{c}{\sin(22.5°)}$$

Generating the Motif

```
//scale factor
let a = 50;

function setup() {
  createCanvas(400, 400);
  angleMode(DEGREES);
  noFill();
  noLoop();
}
function draw() {
  let b = 2 * a * cos(67.5);
  let c = 3 * b * sin(22.5) / sin(112.5);
  push();
    translate(width * 0.5, height * 0.5);
    rotate(45);
    let x0, y0, x1, y1, x2, y2, x3, y3, x4, y4, x5, y5, x6, y6;

    x0 = 0;
    y0 = a;
    x1 = 2 * a * cos(45);
    y1 = 2 * a * sin(45) + a;
    x2 = x1 + b * cos(22.5);
    y2 = y1 - b * sin(22.5);
    x3 = a * cos(45);
    y3 = a * sin(45);
    x4 = x3;
    y4 = y3 + 2 * a;
    x5 = 0;
    y5 = 3 * a;
    x6 = c * cos(22.5);
    y6 = y5 + c * sin(22.5);

    for (let i = 0; i < 4; i++) {
      push();
        rotate(90*i);
          //right side
          beginShape();
          vertex(x0, -y0);
          vertex(x1, -y1);
          vertex(x2, -y2);
          endShape();
          beginShape();
          vertex(x3,-y3);
          vertex(x4,-y4);
          vertex(x5,-y5);
          vertex(x6,-y6);
          endShape();
```

Generating the Motif

```
//left side
push();
    scale(-1,1);
    beginShape();
    vertex(x0, -y0);
    vertex(x1, -y1);
    vertex(x2, -y2);
    endShape();
    beginShape();
    vertex(x3,-y3);
    vertex(x4,-y4);
    vertex(x5,-y5);
    vertex(x6,-y6);
    endShape();
    pop();
  pop();
}
pop();
}
```

Analyzing the Tessellation

Step 2: We need to calculate the dx and dy values in the placement.

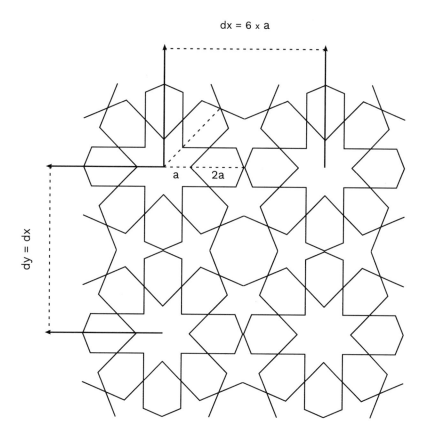

Tessellation Code

```
// Motif class
class Motif {
  constructor(a) {
    this.a = a;
  }
  display() {

    let x0, y0, x1, y1, x2, y2, x3, y3, x4, y4, x5, y5, x6, y6;
    let b = 2 * a * cos(67.5);
    let c = 3 * b * sin(22.5) / sin(112.5);

    x0 = 0;
    y0 = a;
    x1 = 2 * a * cos(45);
    y1 = 2 * a * sin(45) + a;
    x2 = x1 + b * cos(22.5);
    y2 = y1 - b * sin(22.5);
    x3 = a * cos(45);
    y3 = a * sin(45);
    x4 = x3;
    y4 = y3 + 2 * a;
    x5 = 0;
    y5 = 3 * a;
    x6 = c * cos(22.5);
    y6 = y5 + c * sin(22.5);

    push();
      rotate(45);
      for (let i = 0; i < 4; i++) {
        push();
          rotate(90*i);
            //right side
            beginShape();
            vertex(x0, -y0);
            vertex(x1, -y1);
            vertex(x2, -y2);
            endShape();
            beginShape();
            vertex(x3,-y3);
            vertex(x4,-y4);
            vertex(x5,-y5);
            vertex(x6,-y6);
            endShape();
            //left side
            push();
              scale(-1,1);
              beginShape();
              vertex(x0, -y0);
```

Tessellation Code

```
                vertex(x2, -y2);
                endShape();
                beginShape();
                vertex(x3,-y3);
                vertex(x4,-y4);
                vertex(x5,-y5);
                vertex(x6,-y6);
                endShape();
            pop();
        pop();
      }
    pop();

  }
}

//scale factor
let a = 16;
let nRow;
let nCol;
let motif = new Motif(a);
let dx,dy;

function setup() {
  createCanvas(800, 800);
  angleMode(DEGREES);
  noLoop();
  noFill();

  dx = 6 * a;
  dy = dx;

  //approximate the nRow and nCol values
  nRow = ceil(height / dy);
  nCol = ceil(width / dx);
}

function draw() {
  for (let r = 0; r < nRow; r++) {
    for (let c = 0; c < nCol; c++) {
      push();
        translate(c*dx,  r*dy);
        motif.display();
      pop();
    }
  }
}
```

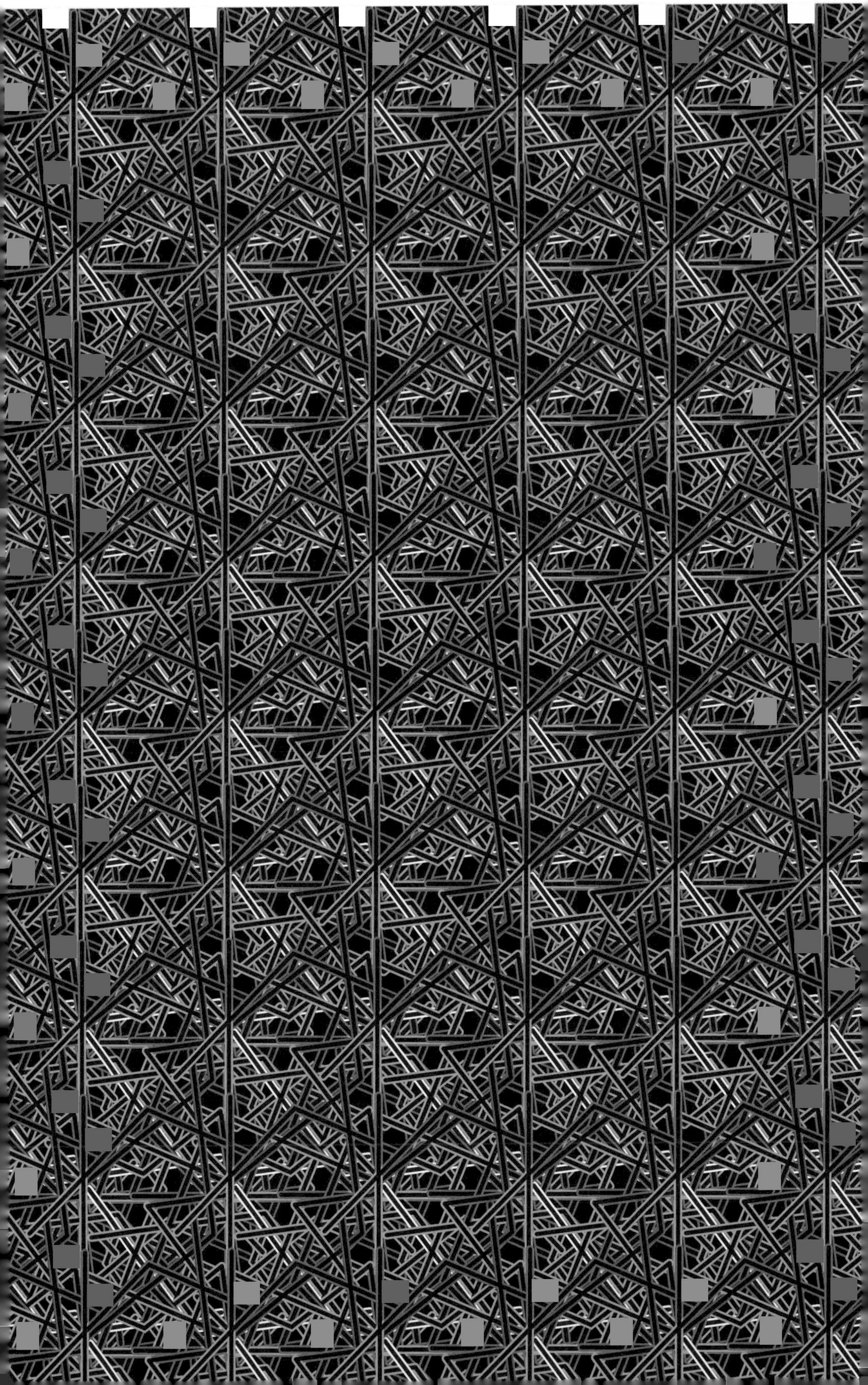

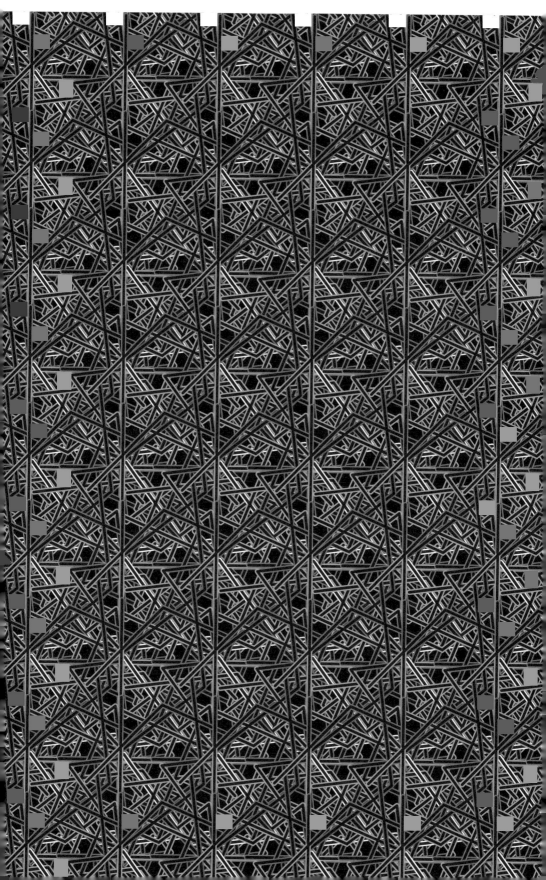

Generating a Geometric Pattern Workflow #20

Observe the geometric pattern, and analyze it to distinguish its constituent repeating motif.

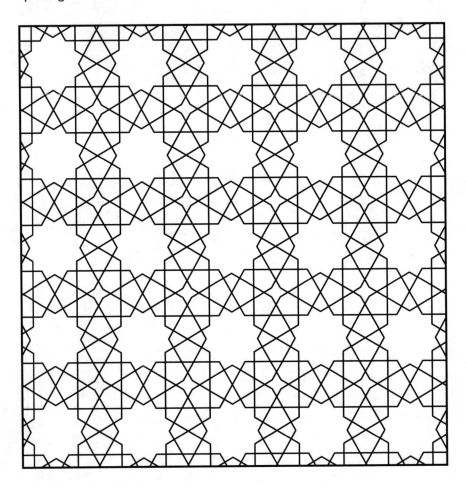

The Motif

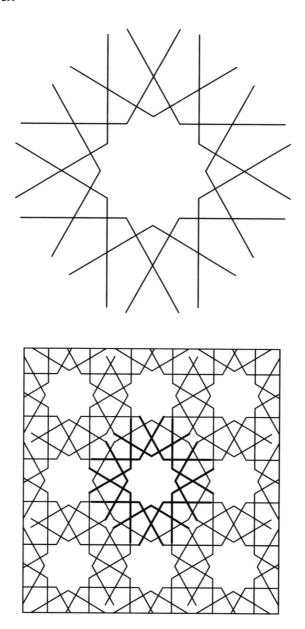

Analyzing the Constructive Elements

There are two types of repeating shapes.

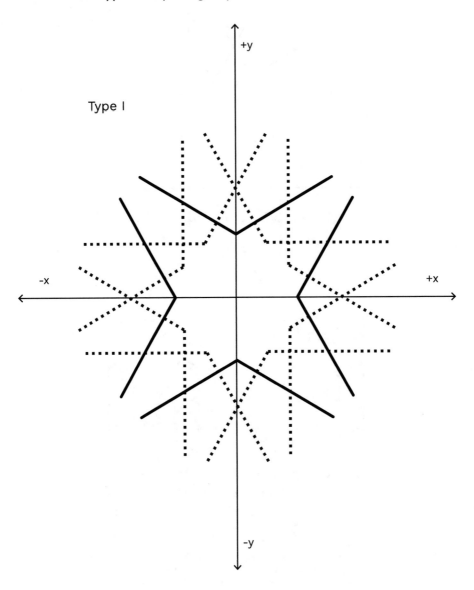

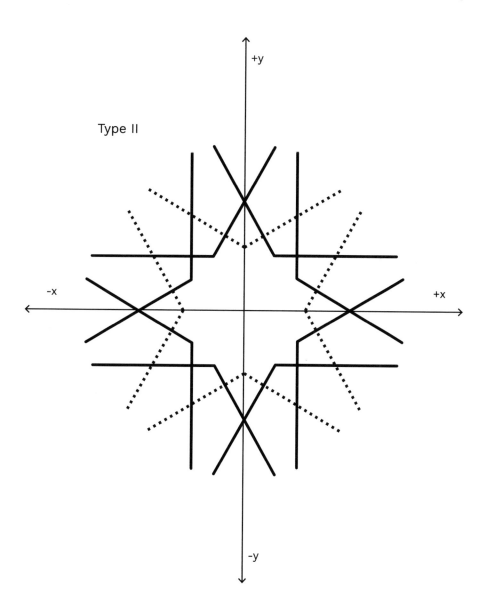

Here type I can be determined with the help of the tessellation.

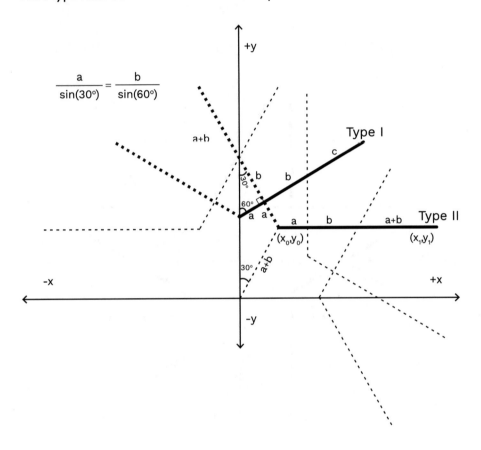

Step 1: Calculate type I's length.

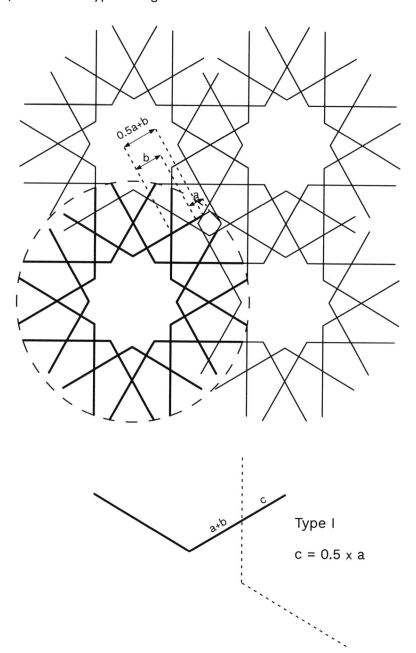

Type I

$c = 0.5 \times a$

Step 2: Let's find the vertex points of the constructive element.

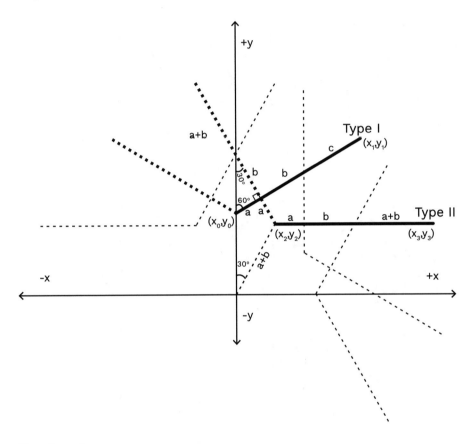

Type I vertices:

$x_0 = 0$

$y_0 = a + b$

$x_1 = (a + 2xb + c) \times \cos(30°)$

$y_1 = (a + 2xb + c) \times \sin(30°) + y_0$

Type II will be drawn like type I and be rotated 30° clockwise.

$x_2 = 0$

$y_2 = a + b$

$x_3 = 2 \times (a + b) \times \cos(30°)$

$y_3 = 2 \times (a + b) \times \sin(30°) + y_2$

Generating the Motif

```
//scale factor
let a = 25;
function setup() {
  createCanvas(400, 400);
  angleMode(DEGREES);
  noLoop();
  noFill();
}

function draw(){
  push();
    translate(width * 0.5, height * 0.5);
    let b = (a * sin(60)) / sin(30);
    let c = 0.5 * a;

    let x0, y0, x1, y1, x2, y2, x3, y3;

    x0 = 0;
    y0 = a + b;
    x1 = (a + 2 * b + c) * cos(30);
    y1 = (a + 2 * b + c) * sin(30) + y0;
    x2 = 0;
    y2 = a + b;
    x3 = 2 * (a + b) * cos(30);
    y3 = 2 * (a + b) * sin(30) + y2;

    for (let i = 0; i < 12; i++) {
      push();
        rotate(30 * i);
        if(i %3 == 0){
          //0,4,8,12
          //shape1
          beginShape();
          vertex(x0, -y0);
          vertex(x1, -y1);
          endShape();
          push();
            scale(-1, 1);
            beginShape();
            vertex(x0, -y0);
            vertex(x1, -y1);
            endShape();
          pop();
        }else{
          //shape2
```

Generating the Motif

```
beginShape();
vertex(x2, -y2);
vertex(x3, -y3);
endShape();
push();
    scale(-1, 1);
    beginShape();
    vertex(x2, -y2);
    vertex(x3, -y3);
    endShape();
pop();
    }
  pop();
  }
pop();
}
```

Analyzing the Tessellation

Step 3: We need to calculate the dx and dy values in the placement.

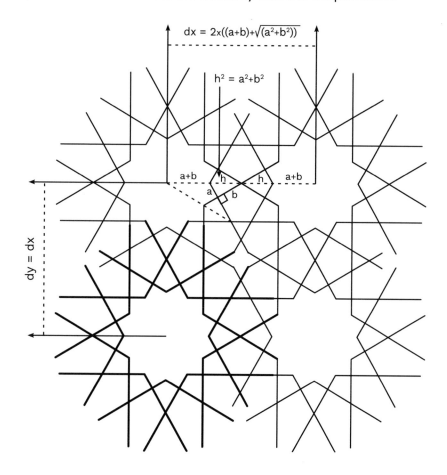

Tessellation Code

```
// Motif class
class Motif {
    constructor(a) {
        this.a = a;
    }
    display() {
        let b = (a * sin(60)) / sin(30);
        let c = a * 0.5 - 0.715; //0.715 pixel based adjustment
        let x0, y0, x1, y1, x2, y2, x3, y3;
        x0 = 0;
        y0 = this.a + b;
        x1 = (this.a + 2 * b + c ) * cos(30);
        y1 = (this.a + 2 * b + c ) * sin(30) + y0;
        x2 = 0;
        y2 = this.a + b;
        x3 = 2 * (this.a + b) * cos(30);
        y3 = 2 * (this.a + b) * sin(30) + y2;

        for (let i = 0; i < 12; i++) {
            push();
                rotate(30 * i);
                if (i % 3 == 0) {
                    //0,4,8,12
                    //shape1
                    beginShape();
                    vertex(x0, -y0);
                    vertex(x1, -y1);
                    endShape();
                    push();
                        scale(-1, 1);
                        beginShape();
                        vertex(x0, -y0);
                        vertex(x1, -y1);
                        endShape();
                    pop();
                }else{
                    //shape2
                    beginShape();
                    vertex(x2, -y2);
                    vertex(x3, -y3);
                    endShape();
                    push();
                        scale(-1, 1);
                        beginShape();
                        vertex(x2, -y2);
                        vertex(x3, -y3);
```

Tessellation Code

```
                        endShape();
                    pop();
                }
            pop();
        }
    }
}

//scale factor
let a = 20;
let motif = new Motif(a);
let nRow;
let nCol;
let xoffset, yoffset;

function setup() {
    createCanvas(800, 800);
    angleMode(DEGREES);
    noFill();
    noLoop();

    let b = (a * sin(60)) / sin(30);

    xoffset = 2 * (a + b) + 2 * sqrt(a * a + b * b);
    yoffset = xoffset;

    //approximate the nRow and nCol values
    nRow = ceil(width / xoffset);
    nCol = ceil(height / yoffset);
}

function draw() {
    for (let c = 0; c < nCol; c++) {
        for (let r = 0; r < nRow; r++) {
            push();
                translate(xoffset * c, yoffset * r);
                motif.display();
            pop();
        }
    }
}
```

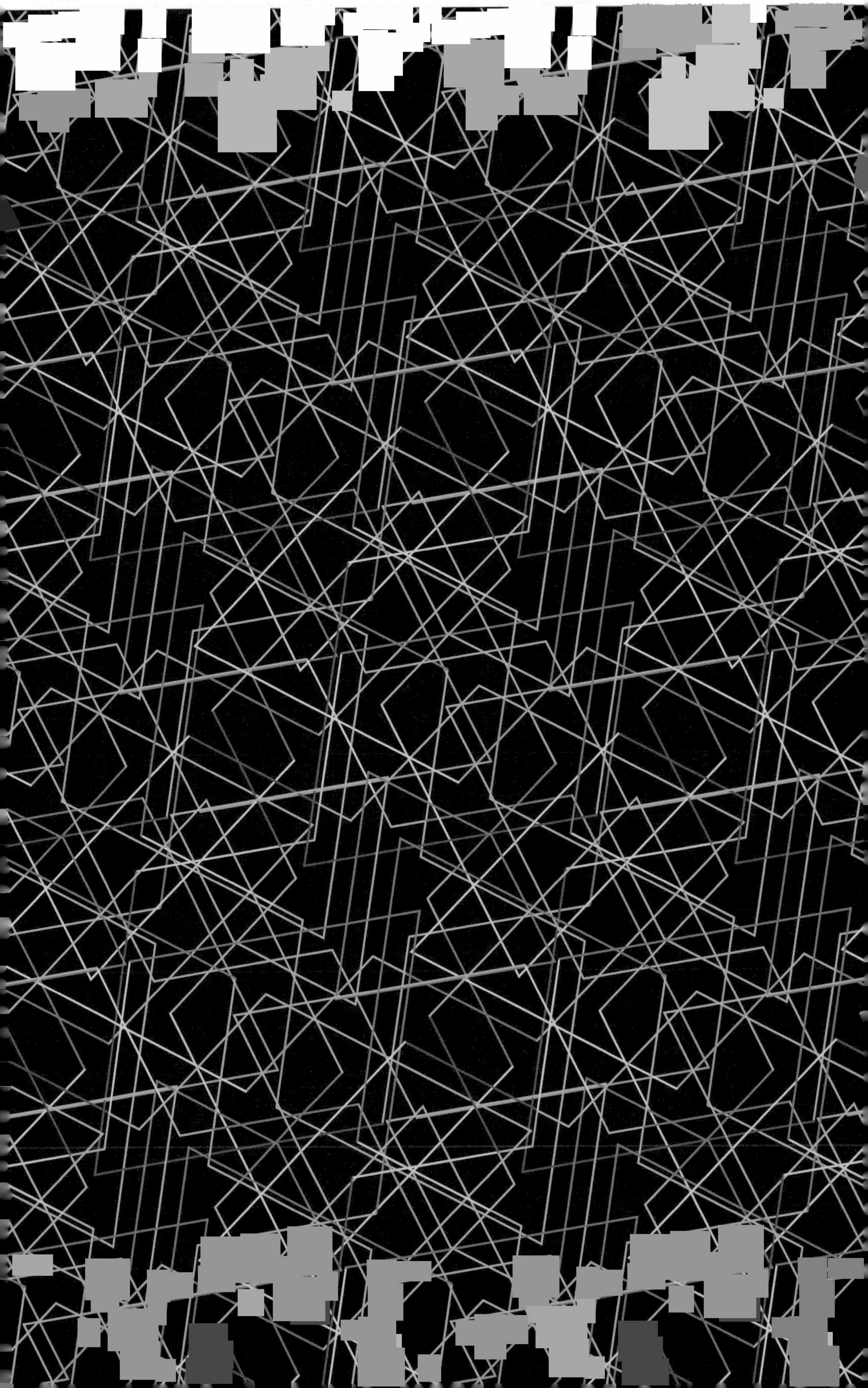

Generating a Geometric Pattern Workflow #21

Observe the geometric pattern, and analyze it to distinguish its constituent repeating motif.

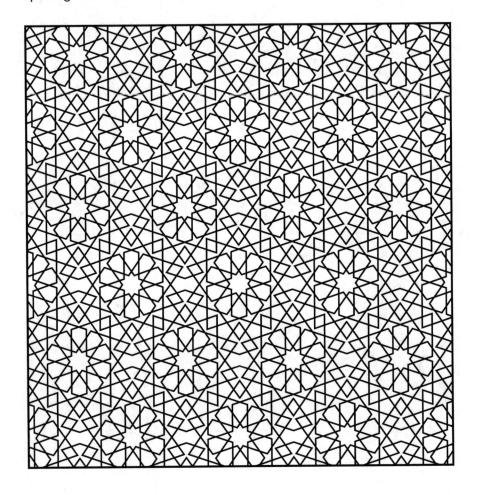

The Motif

Analyzing the Constructive Elements

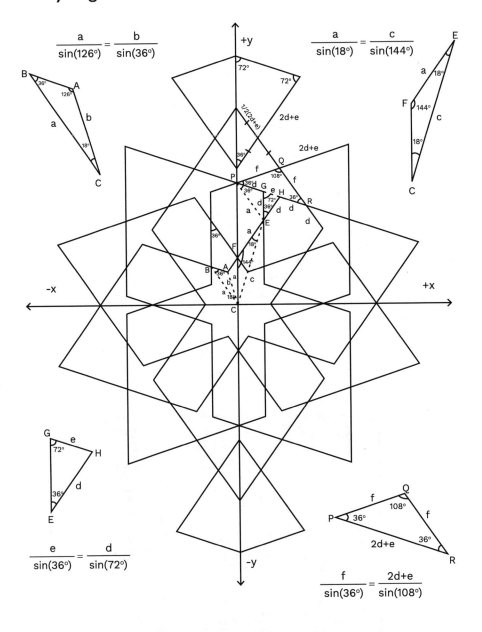

$$\frac{a}{\sin(126°)} = \frac{b}{\sin(36°)}$$

$$\frac{a}{\sin(18°)} = \frac{c}{\sin(144°)}$$

$$\frac{e}{\sin(36°)} = \frac{d}{\sin(72°)}$$

$$\frac{f}{\sin(36°)} = \frac{2d+e}{\sin(108°)}$$

Step 1: Let's focus on the first shape.

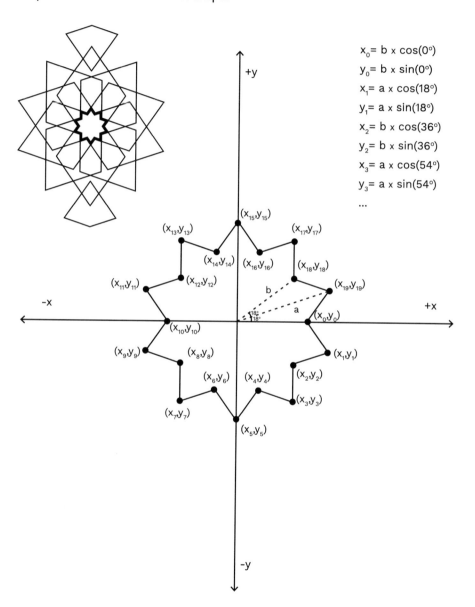

$x_0 = b \times \cos(0°)$

$y_0 = b \times \sin(0°)$

$x_1 = a \times \cos(18°)$

$y_1 = a \times \sin(18°)$

$x_2 = b \times \cos(36°)$

$y_2 = b \times \sin(36°)$

$x_3 = a \times \cos(54°)$

$y_3 = a \times \sin(54°)$

...

Generating the Motif: Shape 1

```
//scale factor
let a = 30;
let b;

function setup() {
  createCanvas(400, 400);
  angleMode(DEGREES);
  noFill();
  noLoop();
  b = a*sin(36)/sin(126);
}

function draw() {
  push();
    translate(width*0.5,height*0.5);
    //Shape 1
    push();
      rotate(18);
      beginShape();
      for(let i = 0; i < 10; i++){
        let x = a * cos(i*36);
        let y = a * sin(i*36);
        vertex(x,y);
        x = b * cos(i*36+18);
        y = b * sin(i*36+18);
        vertex(x,y);
      }
      endShape(CLOSE);
    pop();
  pop();
}
```

Analyzing the Constructive Elements

Step 2: Let's focus on the second shape.

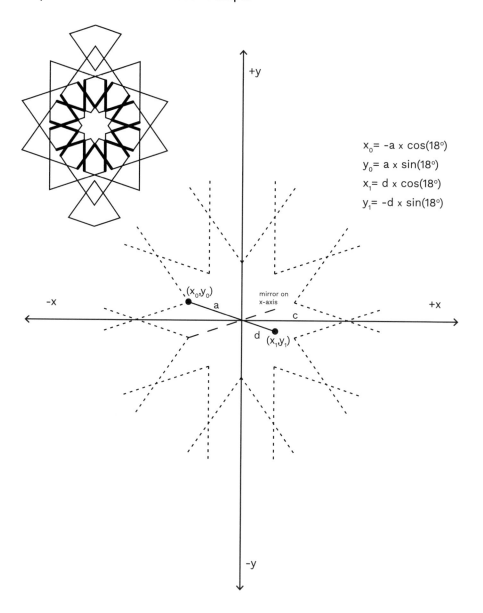

$$x_0 = -a \times \cos(18°)$$
$$y_0 = a \times \sin(18°)$$
$$x_1 = d \times \cos(18°)$$
$$y_1 = -d \times \sin(18°)$$

+y

(x_0, y_0)

a

mirror on
x-axis

c

-x

+x

d

(x_1, y_1)

-y

Generating the Motif: Shape 2

```
//scale factor
let a = 30;
let b,c,d;

function setup() {
  createCanvas(400, 400);
  angleMode(DEGREES);
  noFill();
  noLoop();
  b = a*sin(36)/sin(126);
  c = a*sin(144)/sin(18);
  d = a*sin(36)/sin(108);
}

function draw() {
  push();
    translate(width*0.5,height*0.5);
    //Shape 1
    push();
      rotate(18);
      beginShape();
      for(let i = 0; i < 10; i++){
        let x = a * cos(i*36);
        let y = a * sin(i*36);
        vertex(x,y);
        x = b * cos(i*36+18);
        y = b * sin(i*36+18);
        vertex(x,y);
      }
      endShape(CLOSE);
    pop();

    //Shape 2
    for(let i = 0; i < 10; i++){
      push();
        rotate(i*36);
        translate(c,0);

        beginShape();
        let x0 = -a*cos(18);
        let y0 = a*sin(18);
        let x1 = d*cos(18);
        let y1 = -d*sin(18);
        vertex(x0,y0);
        vertex(x1,y1);
        endShape();
```

Generating the Motif: Shape 2

```
//mirror on x-axis
push();
  scale(1,-1);
  beginShape();
  x0 = -a*cos(18);
  y0 = a*sin(18);
  x1 = d*cos(18);
  y1 = -d*sin(18);
  vertex(x0,y0);
  vertex(x1,y1);
  endShape();
pop();

  pop();
}
pop();
}
```

Analyzing the Constructive Elements

Step 3: Let's focus on the third shape.

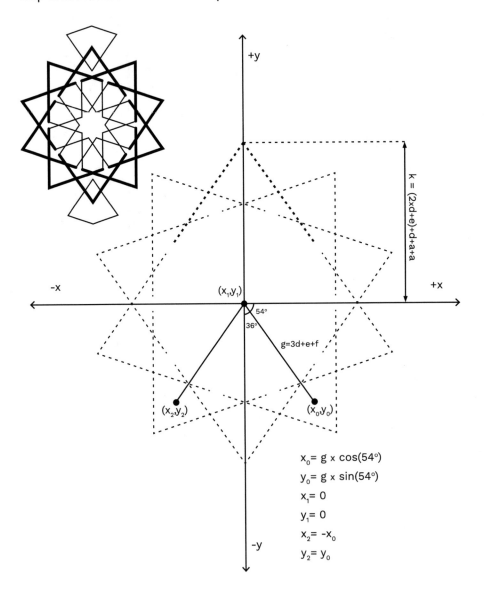

$$x_0 = g \times \cos(54°)$$
$$y_0 = g \times \sin(54°)$$
$$x_1 = 0$$
$$y_1 = 0$$
$$x_2 = -x_0$$
$$y_2 = y_0$$

Generating the Motif: Shape 3

```
//scale factor
let a = 30
let b,c,d,e,f,g,h,j,k;

function setup() {
  createCanvas(400, 400);
  angleMode(DEGREES);
  noFill();
  noLoop();
  b = a*sin(36)/sin(126);
  c = a*sin(144)/sin(18);
  d = a*sin(36)/sin(108);
  e = d*sin(36)/sin(72);
  f = (2*d+e)*sin(36)/sin(108);
  g = 3*d+e+f;
  h = (2*d+e+f)/3;
  j = h *sin(108)/sin(36);
  k = (2*d+e)+d+a+a;
}

function draw() {
  push();
    translate(width*0.5,height*0.5);
    push();
      rotate(18);
      beginShape();
      for(let i = 0; i < 10; i++){
        let x = a * cos(i*36);
        let y = a * sin(i*36);
        vertex(x,y);
        x = b * cos(i*36+18);
        y = b * sin(i*36+18);
        vertex(x,y);
      }
      endShape(CLOSE);
    pop();

    for(let i = 0; i < 10; i++){
      push();
        rotate(i*36);
        translate(c,0);

        beginShape();
        let x0 = -a*cos(18);
        let y0 = a*sin(18);
        let x1 = d*cos(18);
        let y1 = -d*sin(18);
        vertex(x0,y0);
        vertex(x1,y1);
```

Generating the Motif: Shape 3

```
      endShape();

      //mirror on x-axis
      push();
        scale(1,-1);
        beginShape();
        x0 = -a*cos(18);
        y0 = a*sin(18);
        x1 = d*cos(18);
        y1 = -d*sin(18);
        vertex(x0,y0);
        vertex(x1,y1);
        endShape();
      pop();

    pop();
  }

      //outer v
      let x0,y0,x1,y1,x2,y2;
      for(let i = 0; i < 10; i++){
        push();
          rotate(36*i);
          translate(0,-k);
          beginShape();
          x0 = g * cos(54);
          y0 = g * sin(54);
          x1 = 0;
          y1 = 0;
          x2 = -x0;
          y2 = y0;
          vertex(x0,y0);
          vertex(x1,y1);
          vertex(x2,y2);
          endShape();
        pop();
      }
    pop();
  }
```

Analyzing the Constructive Elements

Step 4: Let's focus on the fourth shape.

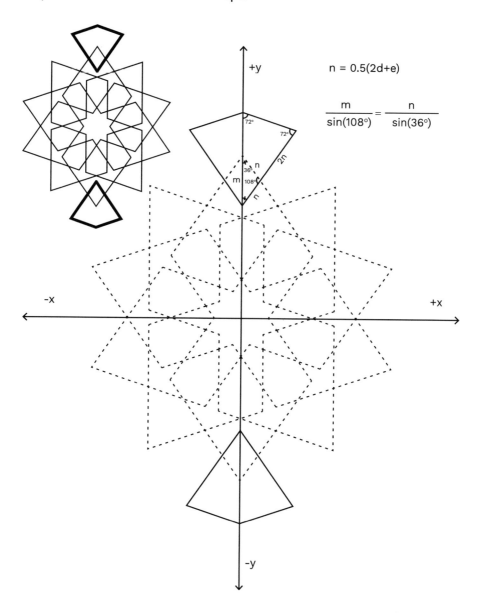

$$n = 0.5(2d+e)$$

$$\frac{m}{\sin(108°)} = \frac{n}{\sin(36°)}$$

Step 5: Let's find the vertex points of the diamond shape.

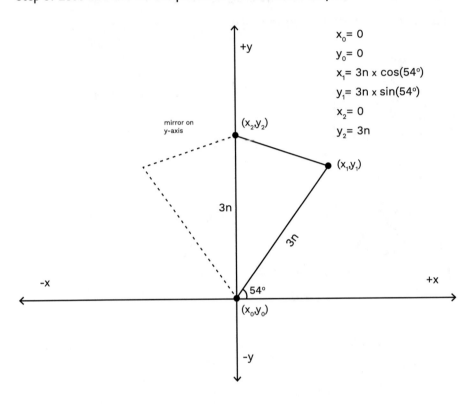

$x_0 = 0$

$y_0 = 0$

$x_1 = 3n \times \cos(54°)$

$y_1 = 3n \times \sin(54°)$

$x_2 = 0$

$y_2 = 3n$

Generating the Motif: Shape 4

```
//scale factor
let a = 30;
let b,c,d,e,f,g,h,j,k,m,n;

function setup() {
  createCanvas(400, 400);
  angleMode(DEGREES);
  noFill();
  noLoop();
  c = a*sin(144)/sin(18);
  b = a*sin(36)/sin(126);

  d = a*sin(36)/sin(108);
  e = d*sin(36)/sin(72);
  f = (2*d+e)*sin(36)/sin(108);
  g = 3*d+e+f;
  h = (2*d+e+f)/3;
  j = h *sin(108)/sin(36);
  k = (2*d+e)+d+a+a;
  m = (2*d+e)*0.5*sin(108)/sin(36);
  n = (2*d+e)*0.5;
}

function draw() {
  push();
    translate(width*0.5,height*0.5);
    //Shape 1
    push();
      rotate(18);
      beginShape();
      for(let i = 0; i < 10; i++){
        let x = a * cos(i*36);
        let y = a * sin(i*36);
        vertex(x,y);
        x = b * cos(i*36+18);
        y = b * sin(i*36+18);
        vertex(x,y);
      }
      endShape(CLOSE);
    pop();

    //Shape 2
    for(let i = 0; i < 10; i++){
      push();
        rotate(i*36);
        translate(c,0);
```

Generating the Motif: Shape 4

```
beginShape();
let x0 = -a*cos(18);
let y0 = a*sin(18);
let x1 = d*cos(18);
let y1 = -d*sin(18);
vertex(x0,y0);
vertex(x1,y1);
endShape();
//mirror on x-axis
push();
   scale(1,-1);
   beginShape();
   x0 = -a*cos(18);
   y0 = a*sin(18);
   x1 = d*cos(18);
   y1 = -d*sin(18);
   vertex(x0,y0);
   vertex(x1,y1);
   endShape();
  pop();
 pop();
}

//Shape 3
let x0,y0,x1,y1,x2,y2;
for(let i = 0; i < 10; i++){
push();
  rotate(36*i);
  translate(0,-k);
  beginShape();
  x0 = g * cos(54);
  y0 = g * sin(54);
  x1 = 0;
  y1 = 0;
  x2 = -x0;
  y2 = y0;
  vertex(x0,y0);
  vertex(x1,y1);
  vertex(x2,y2);
  endShape();
 pop();
 }
```

Generating the Motif: Shape 4

```
//Shape 4
x0 = 0;
y0 = 0;
x1 = 3 * n * cos(54);
y1 = 3 * n * sin(54);
x2 = 0;
y2 = 3 * n;

//one on the top one on the bottom
for(let i=0;i<2;i++){
push();
  rotate(i*180);
  push();
      translate(0,-1*(k-m));
      beginShape();
      vertex(x0,-y0);
      vertex(x1,-y1);
      vertex(x2,-y2);
      endShape();
      push();
        scale(-1,1);
        beginShape();
        vertex(x0,-y0);
        vertex(x1,-y1);
        vertex(x2,-y2);
        endShape();
      pop();
    pop();
  pop();
  }
 pop();
}
```

Analyzing the Tessellation

Step 6: We need to calculate the dx, dy, and doff values in the placement.

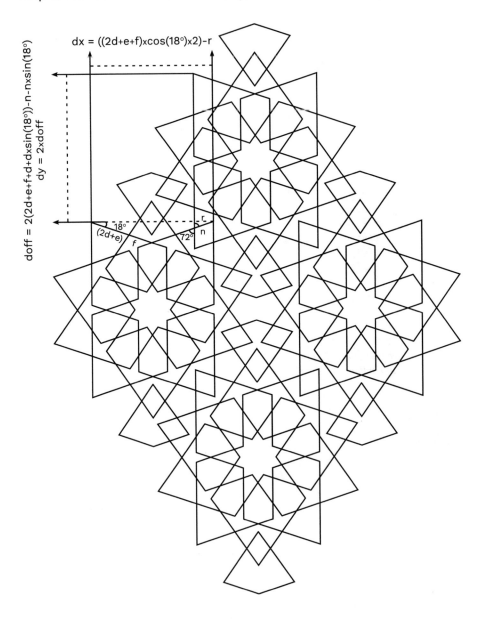

Tessellation Code

```
// Motif class
class Motif {
    constructor(a) {
        this.a = a;
    }

    display() {
        let a, b, c, d, e, f, g, h, j, k, m, n;
        a = this.a;
        c = (a * sin(144)) / sin(18);
        b = (a * sin(36)) / sin(126);
        d = (a * sin(36)) / sin(108);
        e = (d * sin(36)) / sin(72);
        f = ((2 * d + e) * sin(36)) / sin(108);
        g = 3 * d + e + f;
        h = (2 * d + e + f) / 3;
        j = (h * sin(108)) / sin(36)
        k = 2 * d + e + d + a + a;
        m = ((2 * d + e) * 0.5 * sin(108)) / sin(36);
        n = (2 * d + e) * 0.5;

        //Shape 1
        push();
            rotate(18);
            beginShape();
            for(let i = 0; i < 10; i++){
                let x = a * cos(i * 36);
                let y = a * sin(i * 36);
                vertex(x, y);
                x = b * cos(i * 36 + 18);
                y = b * sin(i * 36 + 18);
                vertex(x, y);
            }
            endShape(CLOSE);
        pop();

        //Shape 2
        for(let i = 0; i < 10; i++){
            push();
                rotate(i * 36);
                translate(c, 0);
                beginShape();
                let x0 = -a * cos(18);
                let y0 = a * sin(18);
                let x1 = d * cos(18);
                let y1 = -d * sin(18);
                vertex(x0, y0);
                vertex(x1, y1);
```

Tessellation Code

```
    endShape();
    //mirror on x-axis
    push();
        scale(1, -1);
        beginShape();
        x0 = -a * cos(18);
        y0 = a * sin(18);
        x1 = d * cos(18);
        y1 = -d * sin(18);
        vertex(x0, y0);
        vertex(x1, y1);
        endShape();
    pop();

  pop();
}

//Shape 3
let x0, y0, x1, y1, x2, y2;
for(let i = 0; i < 10; i++){
    push();
        rotate(36 * i);
        translate(0, -k);
        beginShape();
        x0 = g * cos(54);
        y0 = g * sin(54);
        x1 = 0;
        y1 = 0;
        x2 = -x0;
        y2 = y0;
        vertex(x0, y0);
        vertex(x1, y1);
        vertex(x2, y2);
        endShape();
    pop();
}

//Shape 4
x0 = 0;
y0 = 0;
x1 = 3 * n * cos(54);
y1 = 3 * n * sin(54);
x2 = 0;
y2 = 3 * n;
```

Tessellation Code

```
        //one on the top one on the bottom
        for (let i = 0; i < 2; i++) {
            push();
                rotate(i*180);
                push();
                    translate(0, -1*(k-m));
                    beginShape();
                    vertex(x0, -y0);
                    vertex(x1, -y1);
                    vertex(x2, -y2);
                    endShape();
                    push();
                        scale(-1,1);
                        beginShape();
                        vertex(x0,-y0);
                        vertex(x1,-y1);
                        vertex(x2,-y2);
                        endShape();
                    pop();
                pop();
            pop();
        }
    }
}

//scale factor
let a = 25;
let motif = new Motif(a);
let nRow;
let nCol;
let dx, dy, doff;

function setup() {
    createCanvas(800, 800);
    angleMode(DEGREES);
    noFill();
    noLoop();

    let b, c, d, e, f, g, h, j, n, p, r;
    c = (a * sin(144)) / sin(18);
    b = (a * sin(36)) / sin(126);
    d = (a * sin(36)) / sin(108);
    e = (d * sin(36)) / sin(72);
    f = ((2 * d + e) * sin(36)) / sin(108);
    g = 3 * d + e + f;
```

Tessellation Code

```
p = (g * sin(36)) / sin(72);
h = p / 2;
j = (h * sin(108)) / sin(36);
n = (2 * d + e) * 0.5;
p = 2 * (g + d * sin(18));
r = 0.5 * (2 * d + e) * sin(72);
dx = (2 * d + e + f) * cos(18) * 2 - r;
doff = 2 * (2 * d + e + f + d + d * sin(18)) - n - n * sin(18);
dy = 2 * doff;

//approximate the nRow and nCol values
nRow = 1 + ceil(height / dy);
nCol = 1 + ceil(width / dx);
}

function draw() {
    for (let r = 0; r < nRow; r++) {
        for (let c = 0; c < nCol; c++) {
            push();
            if (c % 2 == 1) {
                //columns 1,3,5,7
                translate(0, doff);
            }
            translate(dx * c, dy * r);
            motif.display();
            pop();
        }
    }
}
```

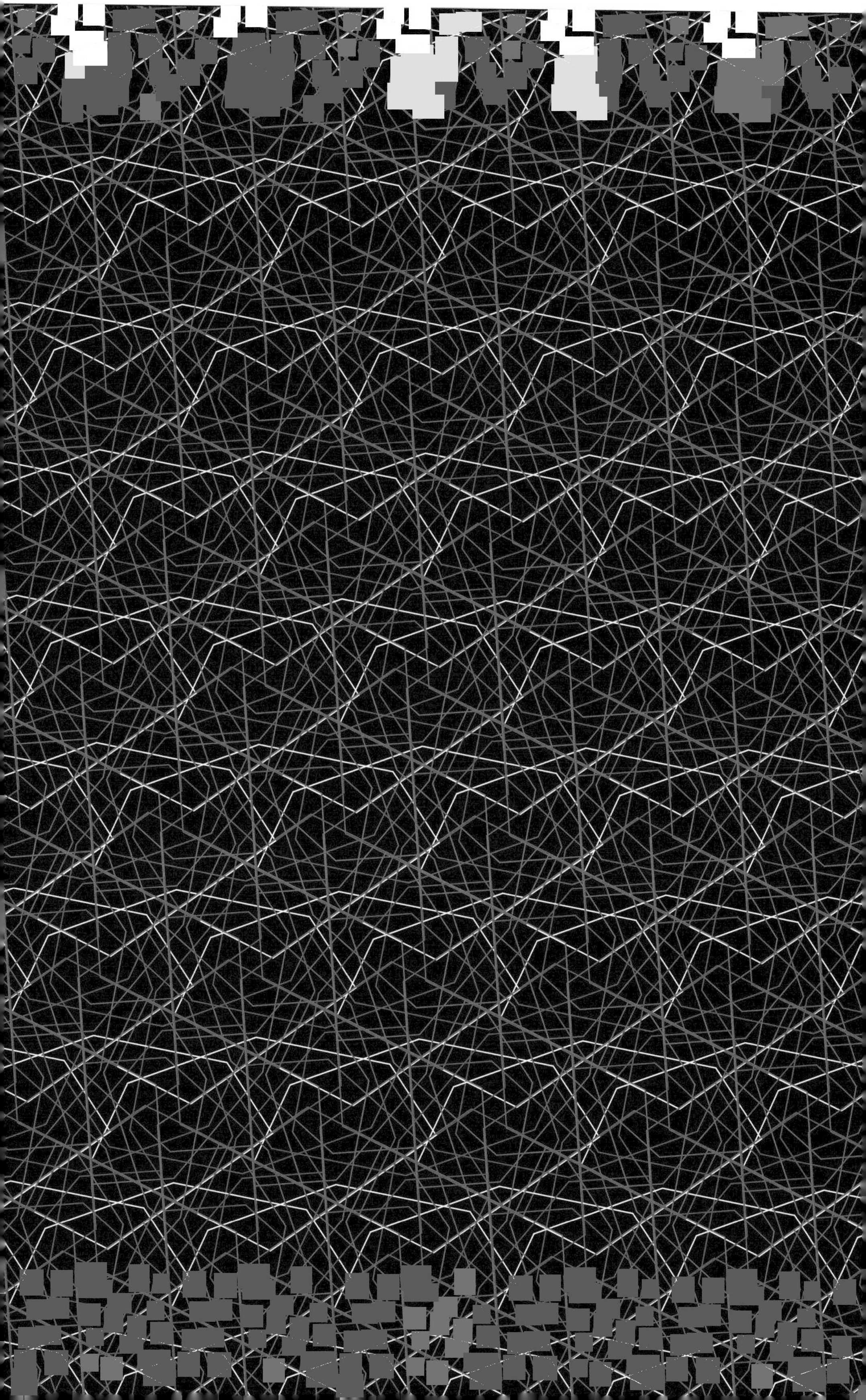

Generating a Geometric Pattern Workflow #22

Observe the geometric pattern, and analyze it to distinguish its constituent repeating motif.

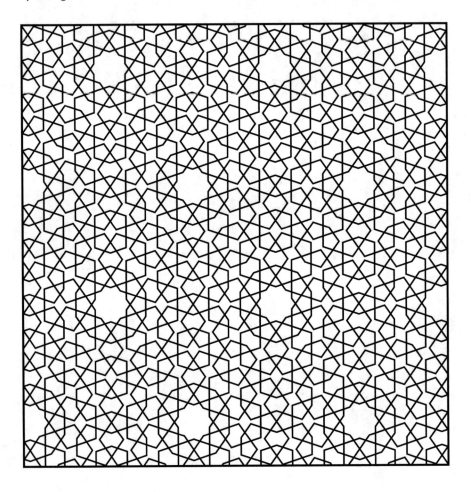

The Motif

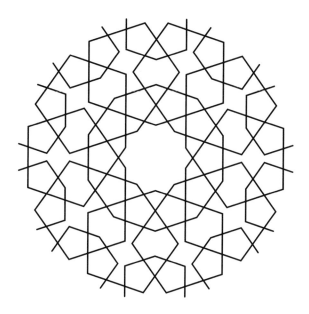

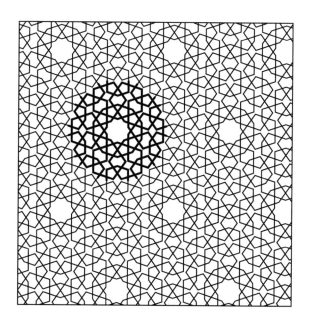

The Filling

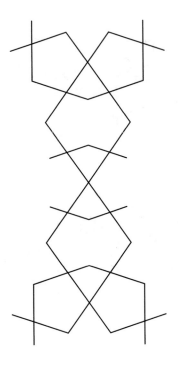

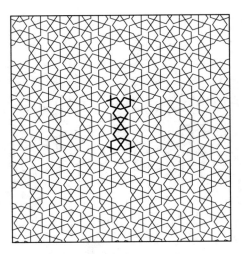

Analyzing the Constructive Elements

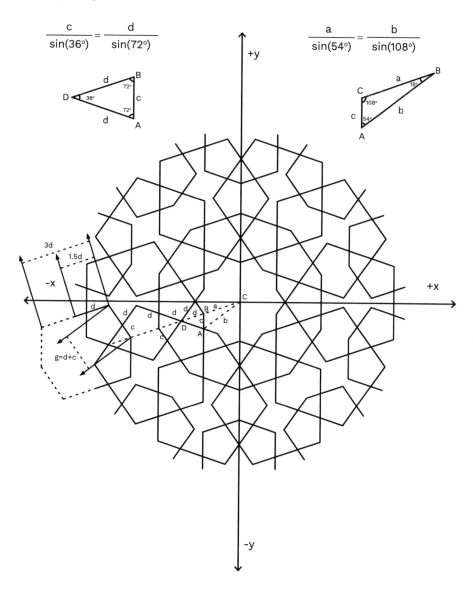

Step 1: Let's focus on the first shape.

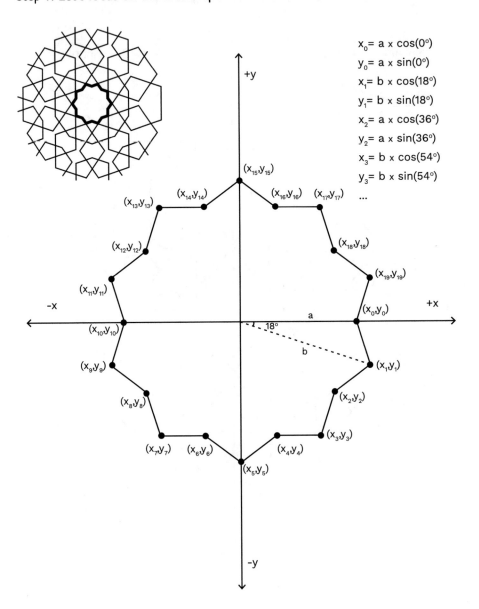

$$x_0 = a \times \cos(0°)$$
$$y_0 = a \times \sin(0°)$$
$$x_1 = b \times \cos(18°)$$
$$y_1 = b \times \sin(18°)$$
$$x_2 = a \times \cos(36°)$$
$$y_2 = a \times \sin(36°)$$
$$x_3 = b \times \cos(54°)$$
$$y_3 = b \times \sin(54°)$$
...

Generating the Motif: Shape 1

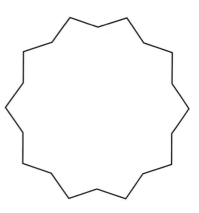

```
//scale factor
let a = 32;
let b;

function setup() {
  createCanvas(400, 400);
  angleMode(DEGREES);
  noFill();
  noLoop();
  b = a * sin(108)/sin(54);
}

function draw() {
  push();
    translate(width*0.5,height*0.5);

    //Shape 1
    let x,y;
    push();
      //orient correctly
      rotate(18);
      beginShape();
      for(let i = 0 ; i < 10; i++){
        x = a * cos(36*i);
        y = a * sin(36*i);
        vertex(x,y);
        x = b * cos(18+36*i);
        y = b * sin(18+36*i);
        vertex(x,y);
      }
      endShape(CLOSE);
    pop();
  pop();
}
```

Analyzing the Constructive Elements

Step 2: Let's focus on the second circular pentagon shape.

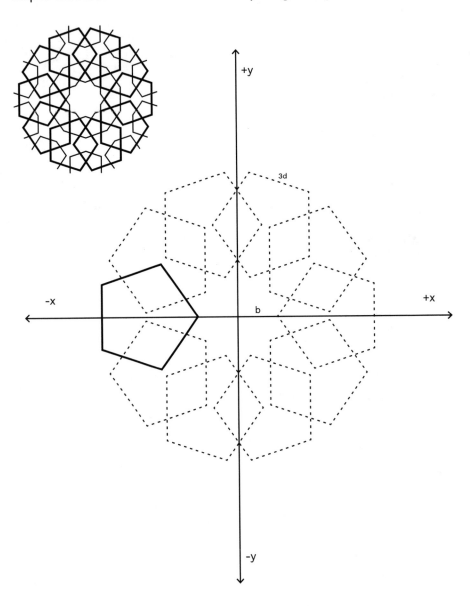

Step 3: Generate the pentagon.

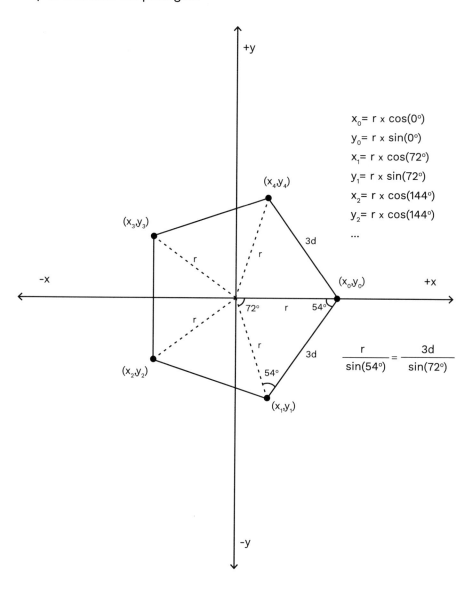

$x_0 = r \times \cos(0°)$
$y_0 = r \times \sin(0°)$
$x_1 = r \times \cos(72°)$
$y_1 = r \times \sin(72°)$
$x_2 = r \times \cos(144°)$
$y_2 = r \times \cos(144°)$

...

$$\frac{r}{\sin(54°)} = \frac{3d}{\sin(72°)}$$

Generating the Motif: Shape 2

One pentagon shape only

```
//scale factor
let a = 32;
let b,c,d;

function setup() {
  createCanvas(400, 400);
  angleMode(DEGREES);
  noFill();
  noLoop();
  b = a * sin(108)/sin(54);
  c = a * sin(18) / sin(54);
  d = c * sin(72) / sin(36);
}

function draw() {
  push();
    translate(width*0.5,height*0.5);
    //Shape 2 pentagon
    let r = 3*d * sin(54)/sin(72);
    beginShape();
    for(let i = 0 ; i < 5; i++){
      x = r * cos(72*i);
      y = r * sin(72*i);
      vertex(x,y);
    }
    endShape(CLOSE);
  pop();
}
```

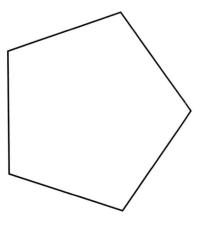

Analyzing the Constructive Elements

Step 4: Apply transformation functions to generate the circular pattern.

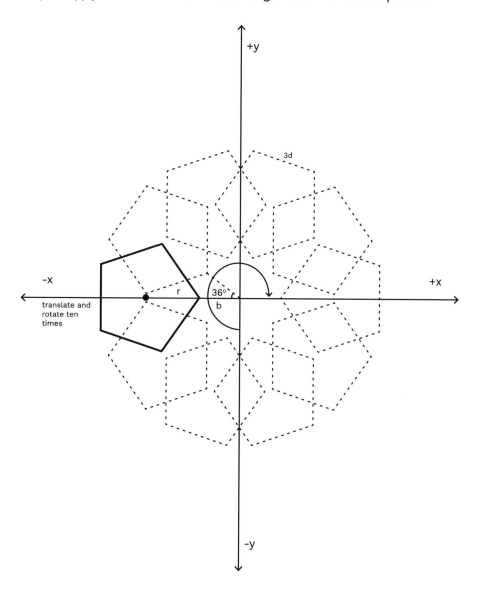

Generating the Motif: Shape 2

Circular pattern with the pentagon shape

```
//scale factor
let a = 32;
let b,c,d;

function setup() {
  createCanvas(400, 400);
  angleMode(DEGREES);
  noFill();
  noLoop();
  b = a * sin(108)/sin(54);
  c = a * sin(18) / sin(54);
  d = c * sin(72) / sin(36);
}

function draw() {
  push();
    translate(width*0.5,height*0.5);

    //Shape 2 pentagon in a circular pattern
    let r = 3*d * sin(54)/sin(72);
    for(let i = 0 ; i < 10; i++){
      push();
        rotate(36*i);
        translate(-(b+r),0);
        beginShape();
        for(let i = 0 ; i < 5; i++){
          x = r * cos(72*i);
          y = r * sin(72*i);
          vertex(x,y);
        }
        endShape(CLOSE);
      pop();
    }
  pop();
}
```

Analyzing the Constructive Elements

Step 5: Let's focus on the decagon shape.

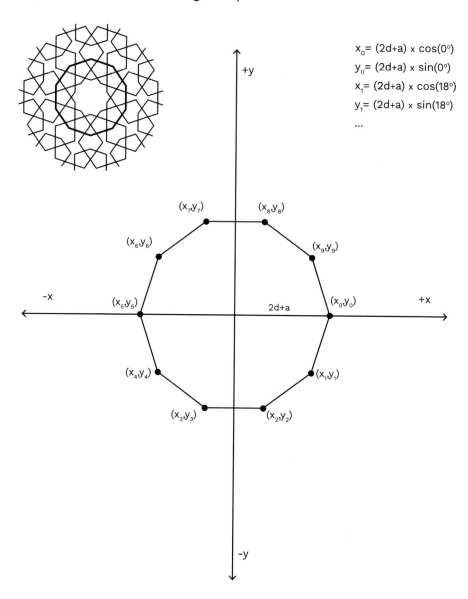

$x_0 = (2d+a) \times \cos(0°)$
$y_0 = (2d+a) \times \sin(0°)$
$x_1 = (2d+a) \times \cos(18°)$
$y_1 = (2d+a) \times \sin(18°)$

...

Generating the Motif: Shape 3

```
//scale factor
let a = 32;
let b,c,d;

function setup() {
  createCanvas(400, 400);
  angleMode(DEGREES);
  noFill();
  noLoop();
  b = a * sin(108)/sin(54);
  c = a * sin(18) / sin(54);
  d = c * sin(72) / sin(36);
}

function draw() {
  push();
    translate(width*0.5,height*0.5);

    //Shape - decagon
    push();
        rotate(18);
        beginShape();
        for(let i = 0 ; i < 10; i++){
          let x0 = (2*d+a) * cos(36*i);
          let y0 = (2*d+a) * sin(36*i);
          vertex(x0,y0);
        }
        endShape(CLOSE);
    pop();

  pop();
}
```

Analyzing the Constructive Elements

Step 6: Let's focus on the surrounding u-type shapes.

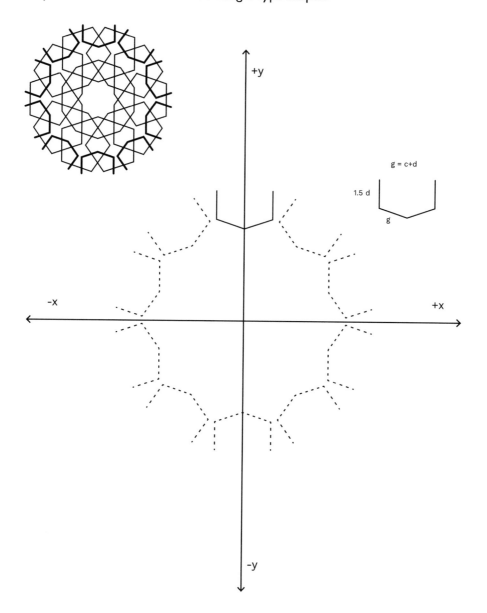

Step 7: Let's find the vertex points of the u-type shape.

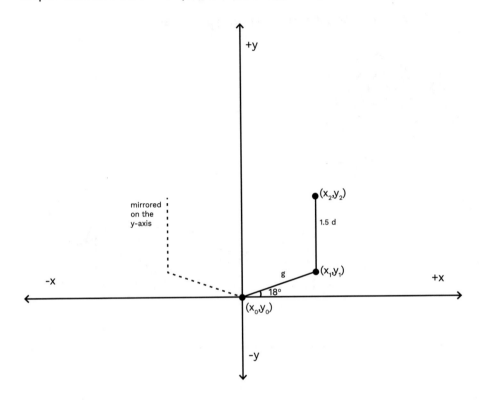

Generating the Motif: Shape 4

One u-type shape only

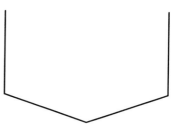

```
//scale factor
let a = 32;
let b,c,d;

function setup() {
  createCanvas(400, 400);
  angleMode(DEGREES);
  noFill();
  noLoop();
  b = a * sin(108)/sin(54);
  c = a * sin(18) / sin(54);
  d = c * sin(72) / sin(36);
}

function draw() {
  push();
    translate(width*0.5,height*0.5);

    //Shape 4 the u-type shape
    let e = 2*d*sin(108)/sin(36);
    let f = e+a;
    let g = d+c;
    translate(0,-f);
    let x0,y0,x1,y1,x2,y2,x3,y3;

    x0 = 0;
    y0 = 0;
    x1 = g*cos(18);
    y1 = g*sin(18);
    x2 = x1;
    y2 = y1+1.5*d;

    beginShape();
    vertex(x0,-y0);
    vertex(x1,-y1);
    vertex(x2,-y2);
    endShape();
    //mirror
    push();
      scale(-1,1);
      beginShape();
      vertex(x0,-y0);
      vertex(x1,-y1);
      vertex(x2,-y2);
      endShape();
    pop();

  pop();
```

Analyzing the Constructive Elements

Step 8: Apply transformation functions to generate the circular pattern made with u-type shapes.

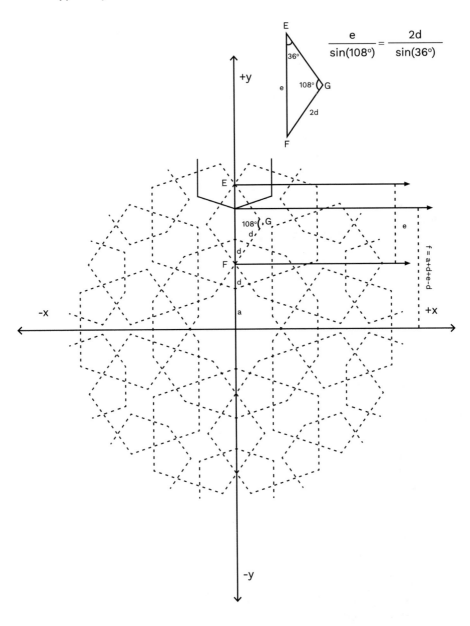

$$\frac{e}{\sin(108°)} = \frac{2d}{\sin(36°)}$$

$f = a+d+e-d$

Generating the Motif: Shape 4

Circular pattern with the u-type shape

```
//scale factor
let a = 32;
let b,c,d;

function setup() {
  createCanvas(400, 400);
  angleMode(DEGREES);
  noFill();
  noLoop();
  b = a * sin(108)/sin(54);
  c = a * sin(18) / sin(54);
  d = c * sin(72) / sin(36);
}

function draw() {
  push();
    translate(width*0.5,height*0.5);

    //Shape 4 circular pattern with the u-type shape
    for(let i = 0 ; i < 10; i++){
    push();
      rotate(i*36);
      let e = 2*d*sin(108)/sin(36);
      let f = e+a;
      let g = d+c;
      translate(0,-f);
      let x0,y0,x1,y1,x2,y2,x3,y3;

      x0 = 0;
      y0 = 0;
      x1 = g*cos(18);
      y1 = g*sin(18);
      x2 = x1;
      y2 = y1+1.5*d;

      beginShape();
      vertex(x0,-y0);
      vertex(x1,-y1);
      vertex(x2,-y2);
      endShape();
```

Generating the Motif: Shape 4

```
//mirror
push();
  scale(-1,1);
  beginShape();
  vertex(x0,-y0);
  vertex(x1,-y1);
  vertex(x2,-y2);
  endShape();
pop();

  pop();
}
pop();
}
```

Generating the Motif

Combining the shapes together

```
//scale factor
let a = 32;
let b,c,d;

function setup() {
  createCanvas(400, 400);
  angleMode(DEGREES);
  noFill();
  noLoop();
  b = a * sin(108)/sin(54);
  c = a * sin(18) / sin(54);
  d = c * sin(72) / sin(36);
}

function draw() {
  push();
    translate(width*0.5,height*0.5);

    //Shape 1 - ten sided star
    let x,y;
    push();
      //orient correctly
      rotate(18);
      beginShape();
      for(let i = 0 ; i < 10; i++){
        x = a * cos(36*i);
        y = a * sin(36*i);
        vertex(x,y);
        x = b * cos(18+36*i);
        y = b * sin(18+36*i);
        vertex(x,y);
      }
      endShape(CLOSE);
    pop();
```

Generating the Motif

Combining the shapes together

```
//Shape 2 - pentagon in a circular pattern
let r = 3*d * sin(54)/sin(72);
for(let i = 0 ; i < 10; i++){
  push();
    rotate(36*i);
    translate(-(b+r),0);
    beginShape();
    for(let i = 0 ; i < 5; i++){
      x = r * cos(72*i);
      y = r * sin(72*i);
      vertex(x,y);
    }
    endShape(CLOSE);
  pop();
}

//Shape 3 - decagon
push();
  rotate(18);
  beginShape();
  for(let i = 0 ; i < 10; i++){
    let x0 = (2*d+a) * cos(36*i);
    let y0 = (2*d+a) * sin(36*i);
    vertex(x0,y0);
  }
  endShape(CLOSE);
pop();

//Shape 4 circular pattern with the u-type shape
for(let i = 0 ; i < 10; i++){
push();
  rotate(i*36);
  let e = 2*d*sin(108)/sin(36);
  let f = e+a;
  let g = d+c;
  translate(0,-f);
  let x0,y0,x1,y1,x2,y2,x3,y3;

  x0 = 0;
  y0 = 0;
  x1 = g*cos(18);
  y1 = g*sin(18);
  x2 = x1;
  y2 = y1+1.5*d;
```

Generating the Motif

Combining the shapes together

```
beginShape();
vertex(x0,-y0);
vertex(x1,-y1);
vertex(x2,-y2);
endShape();
//mirror
push();
    scale(-1,1);
    beginShape();
    vertex(x0,-y0);
    vertex(x1,-y1);
    vertex(x2,-y2);
    endShape();
pop();

pop();
}
pop();
}
```

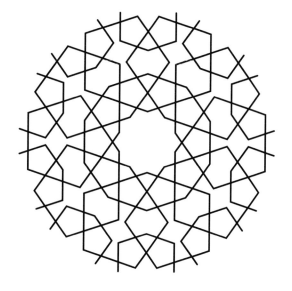

Analyzing the Filling Element

Step 9: Let's find the vertex points of the filling shape.

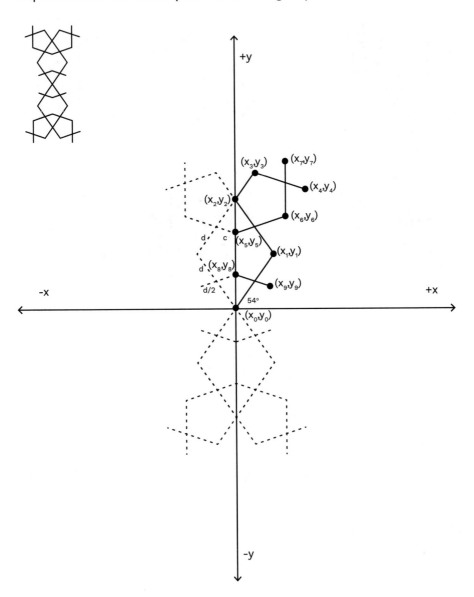

Generating the Filling

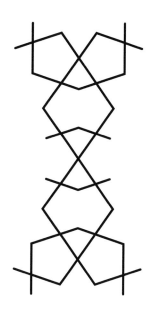

```
//scale factor
let a = 32;

function setup() {
    createCanvas(400, 400);
    angleMode(DEGREES);
    noFill();
    noLoop();
}

function draw() {
    push();
        translate(width * 0.5, height * 0.5);
        let c, d, e;
        c = (a * sin(18)) / sin(54);
        d = (c * sin(72)) / sin(36);
        e = ((d + d) * sin(108)) / sin(36);

        let x0,y0,x1,y1,x2,y2,x3,y3,x4,y4;
        let x5,y5,x6,y6,x7,y7,x8,y8,x9,y9;
        x0 = 0;
        y0 = 0;
        x1 = (d + d) * cos(54);
        y1 = (d + d) * sin(54);
        x2 = 0;
        y2 = e;
        x3 = d * cos(54);
        y3 = y2 + d * sin(54);
        x4 = x3 + (d + c) * cos(18);
        y4 = y3 - (d + c) * sin(18);
        x5 = 0;
        y5 = e - d;
        x6 = (c + d) * cos(18);
        y6 = y5 + (c + d) * sin(18);
        x7 = x6;
        y7 = y6 + (c + d);
        x8 = 0;
        y8 = d;
        x9 = (c + d * 0.5) * cos(18);
        y9 = y8 - (c + d * 0.5) * sin(18);
```

```
push();
    for (let n = -1; n < 2; n += 2) {
        push();
            scale(1, n);
            for (let m = -1; m < 2; m += 2) {
                push();
                    scale(m, 1);
                    beginShape();
                    vertex(x0, -y0);
                    vertex(x1, -y1);
                    vertex(x2, -y2);
                    vertex(x3, -y3);
                    vertex(x4, -y4);
                    endShape();
                    beginShape();
                    vertex(x5, -y5);
                    vertex(x6, -y6);
                    vertex(x7, -y7);
                    endShape();
                    beginShape();
                    vertex(x8, -y8);
                    vertex(x9, -y9);
                    endShape();
                pop();
            }
        pop();
    }
pop();
}
```

Analyzing the Tessellation

Step 10: We need to calculate the dx, dy, and doff values of the motif and the filling in the placement.

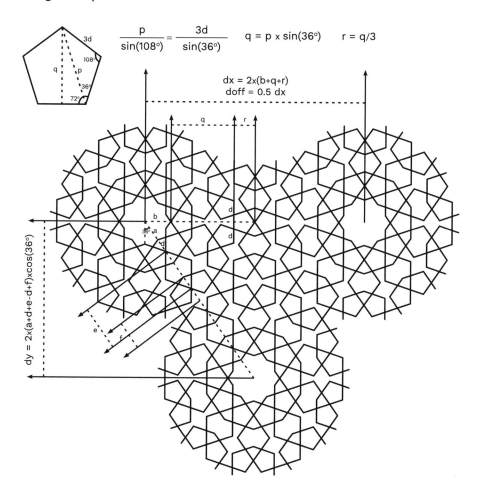

$$\frac{p}{\sin(108°)} = \frac{3d}{\sin(36°)} \qquad q = p \times \sin(36°) \qquad r = q/3$$

$$dx = 2 \times (b+q+r)$$
$$doff = 0.5\ dx$$

$$dy = 2 \times (a+d+e-d+f) \times \cos(36°)$$

Tessellation Code

```
//Filling class
class Filling {
  constructor(a) {
    this.a = a;
  }

  display() {
    let c,d,e;
    c - this.a * sin(18) / sin(54);
    d = c * sin(72) / sin(36);
    e = (d+d) * sin(108)/ sin(36)

    let x0,y0,x1,y1,x2,y2,x3,y3,x4,y4,x5,y5,x6,y6,x7,y7,x8,y8,x9,y9;
    x0 = 0;
    y0 = 0;
    x1 = (d+d) * cos(54);
    y1 = (d+d) * sin(54);
    x2 = 0;
    y2 = e;
    x3 = d * cos(54);
    y3 = y2 + d * sin(54);
    x4 = x3+(d+c) * cos(18);
    y4 = y3-(d+c) * sin(18);
    x5 = 0;
    y5 = e-d;
    x6 = (c+d) * cos(18);
    y6 = y5+(c+d) * sin(18);
    x7 = x6;
    y7 = y6 + (c+d);
    x8 = 0;
    y8 = d;
    x9 = (c+d*0.5)*cos(18);
    y9 = y8 - (c+d*0.5)*sin(18);

    push();
      for(let n = -1;n<2;n+=2){
        push();
          scale(1,n);
          for(let m = -1;m<2;m+=2){
            push();
              scale(m,1);
              beginShape();
              vertex(x0,-y0);
              vertex(x1,-y1);
              vertex(x2,-y2);
              vertex(x3,-y3);
              vertex(x4,-y4);
              endShape();
```

Tessellation Code

```
            beginShape();
            vertex(x5,-y5);
            vertex(x6,-y6);
            vertex(x7,-y7);
            endShape();

            beginShape();
            vertex(x8,-y8);
            vertex(x9,-y9);
            endShape();
          pop();
        }
      pop();
    }
  pop();
  }
}

//Motif class
class Motif {
  constructor(a) {
    this.a = a;
  }

  display() {

    let a = this.a;
    let b,c,d;

    b = a * sin(108)/sin(54);
    c = a * sin(18) / sin(54);
    d = c * sin(72) / sin(36);

    //Shape 1 - ten sided star
    let x,y;
    push();
      //orient correctly
      rotate(18);
      beginShape();
      for(let i = 0 ; i < 10; i++){
        x = a * cos(36*i);
        y = a * sin(36*i);
        vertex(x,y);
        x = b * cos(18+36*i);
        y = b * sin(18+36*i);
        vertex(x,y);
      }
      endShape(CLOSE);
    pop();
```

Tessellation Code

```
//Shape 2 - pentagon in a circular pattern
let r = 3*d * sin(54)/sin(72);
for(let i = 0 ; i < 10; i++){
push();
  rotate(36*i);
  translate(-(b+r),0);
  beginShape();
  for(let i = 0 ; i < 5; i++){
    x = r * cos(72*i);
    y = r * sin(72*i);
    vertex(x,y);
  }
  endShape(CLOSE);
pop();
}

//Shape 3 - decagon
push();
  rotate(18);
  beginShape();
  for(let i = 0 ; i < 10; i++){
    let x0 = (2*d+a) * cos(36*i);
    let y0 = (2*d+a) * sin(36*i);
    vertex(x0,y0);
  }
  endShape(CLOSE);
pop();

//Shape 4 circular pattern with the u-type shape
for(let i = 0 ; i < 10; i++){
  push();
    rotate(i*36);
    let e = 2*d*sin(108)/sin(36);
    let f = e+a;
    let g = d+c;
    translate(0,-f);
    let x0,y0,x1,y1,x2,y2,x3,y3;

    x0 = 0;
    y0 = 0;
    x1 = g*cos(18);
    y1 = g*sin(18);
    x2 = x1;
    y2 = y1+1.5*d;
```

Tessellation Code

```
        beginShape();
        vertex(x0,-y0);
        vertex(x1,-y1);
        vertex(x2,-y2);
        endShape();
        //mirror
        push();
            scale(-1,1);
            beginShape();
            vertex(x0,-y0);
            vertex(x1,-y1);
            vertex(x2,-y2);
            endShape();
        pop();

      pop();
        }
      }
}

//scale factor
let a = 25;
let b, c, d, p, q, r;

let motif = new Motif(a);
let filling = new Filling(a);

let dx,dy,doff;
let nRow;
let nCol;

function setup() {
  createCanvas(800, 800);
  angleMode(DEGREES);
  noFill();
  noLoop();

  b = a * sin(108)/sin(54);
  c = a * sin(18) / sin(54);
  d = c * sin(72) / sin(36);

  p = 3*d*sin(108)/sin(36);
  q = p*sin(72); //penta height
  r = q/3;

  dx = 2*(q + b + r);
  doff = dx * 0.5;
```

Tessellation Code

```
let e = 2*d*sin(108)/sin(36);
let f = (d+c)*cos(72)+1.5*d;
let g = e+a+f;

dy = 2*g*cos(36);

//approximate the nRow and nCol values
nRow = 1+ceil(height / dy);
nCol = 1+ceil(width / dx);
}

function draw() {
    for (let r = 0; r < nRow; r++) {
      for (let c = 0; c < nCol; c++) {
        push();
          if(r%2==1){
            //rows 1,3,5,7
            translate(doff,0);
          }
          translate(dx * c,  dy * r);
          motif.display();
        pop();

        push();
          if(r%2==0){
            //columns 1,3,5,7
            translate(doff,0);
          }
          translate(dx * c,  dy * r);
          filling.display();
        pop();
      }
    }
}
```

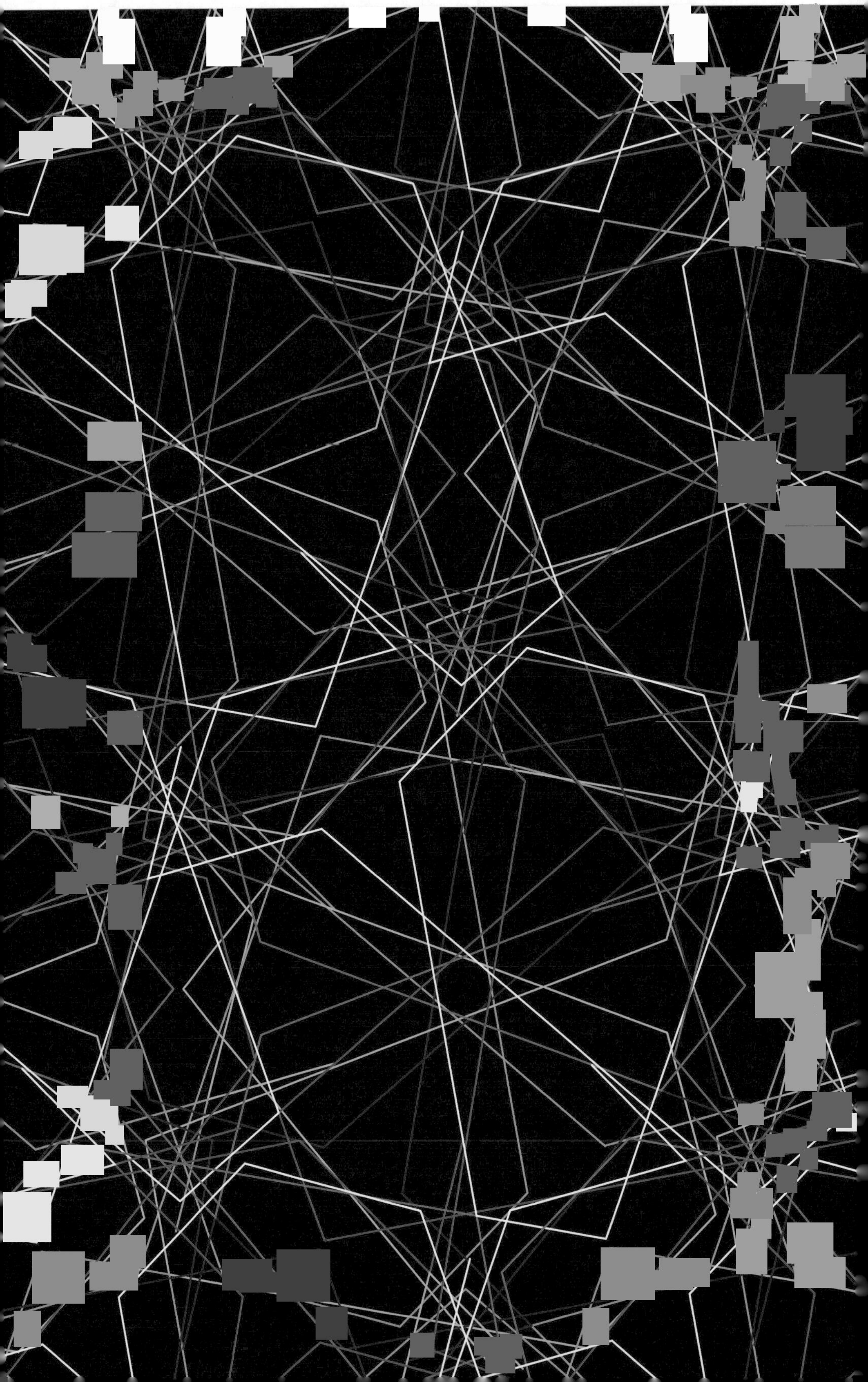

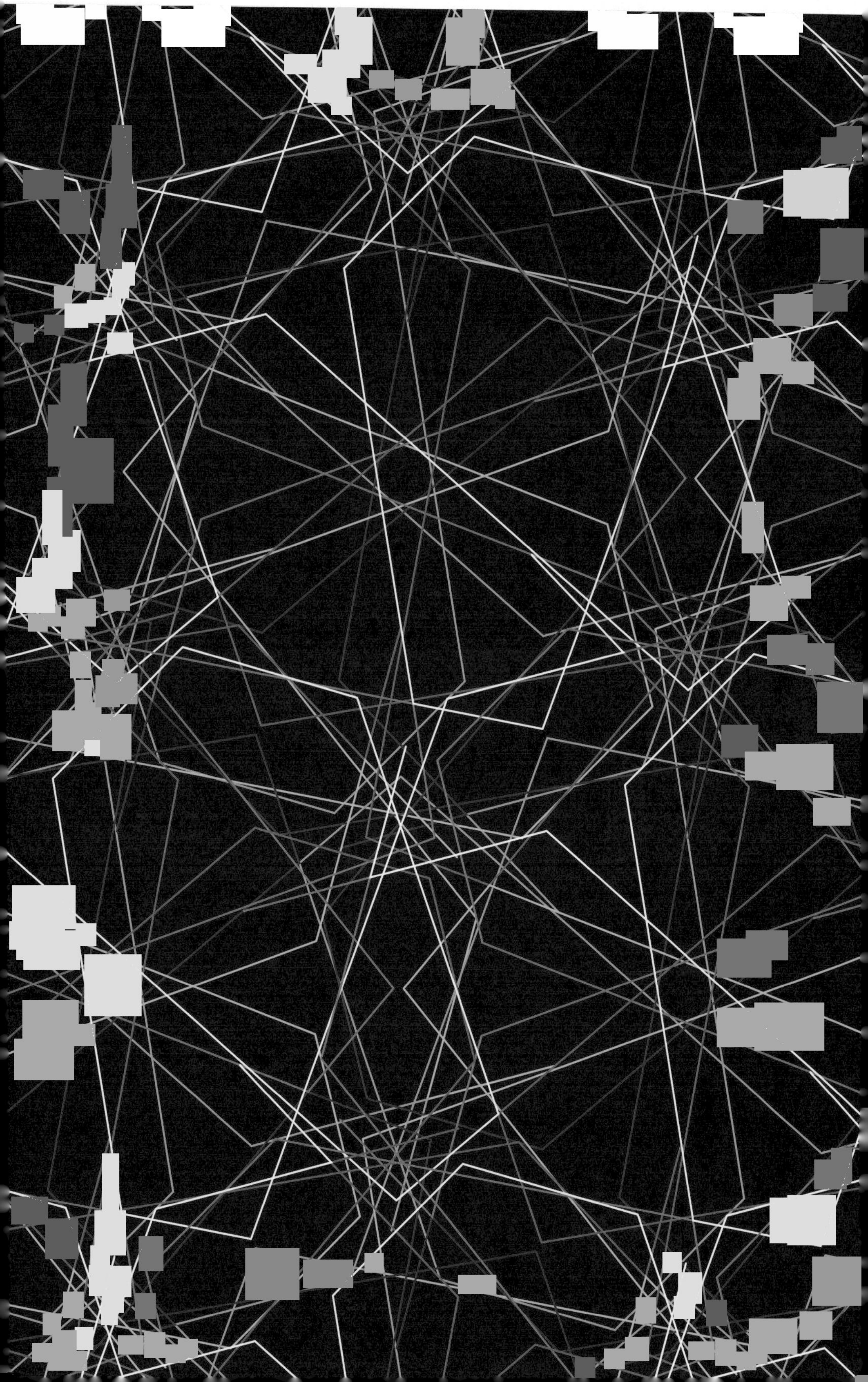

Generating a Geometric Pattern Workflow #23

Observe the geometric pattern, and analyze it to distinguish its constituent repeating motif.

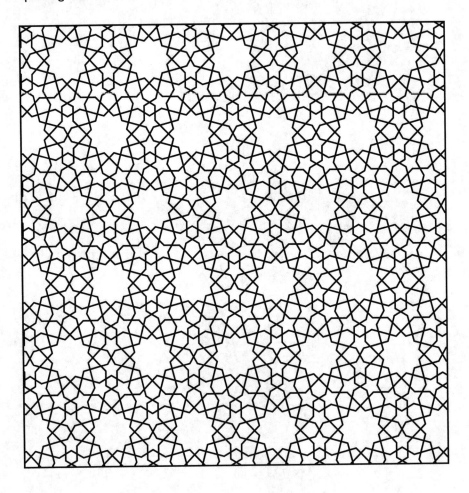

The Motif

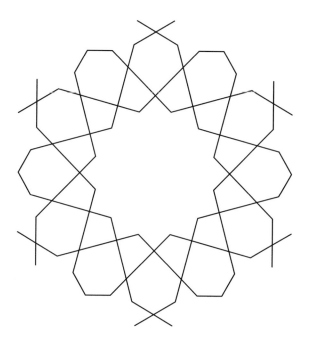

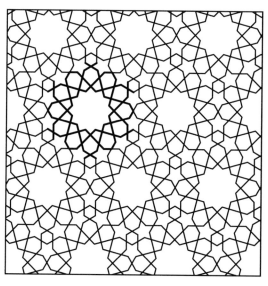

Analyzing the Constructive Elements

Step 1: Let's find the vertex points of the constructive element.

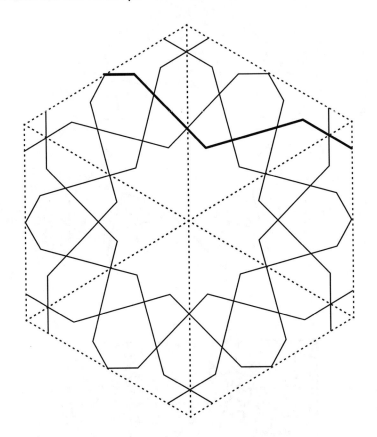

Step 2: Let's find the vertex points of the constructive triangle.

$$\frac{a}{\sin(45°)} = \frac{b}{\sin(120°)} = \frac{c}{\sin(15°)}$$

$$\frac{c}{\sin(30°)} = \frac{d}{\sin(90°)} = \frac{e}{\sin(60°)}$$

$f = (e+c) \times \sin(45°)$

$h = ((b+f) \times \sin(30°) / \sin(60°)) - f$

$k = (b+f) / \sin(60)$

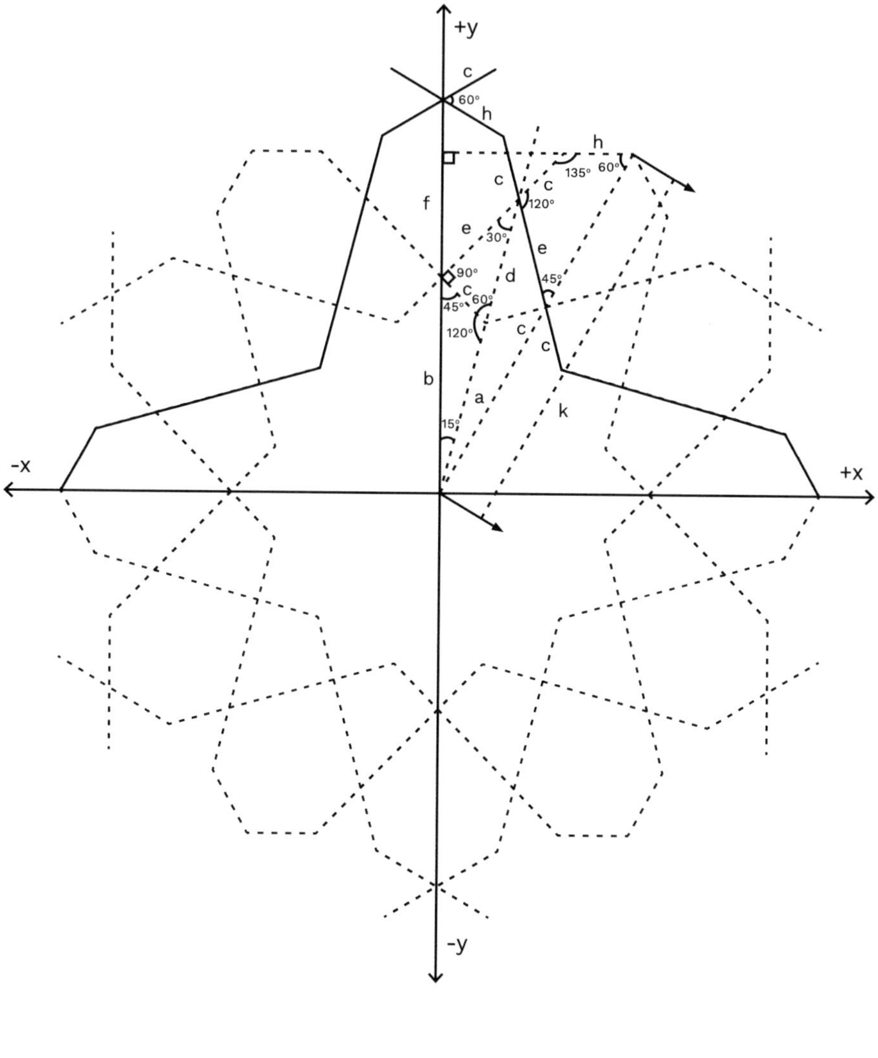

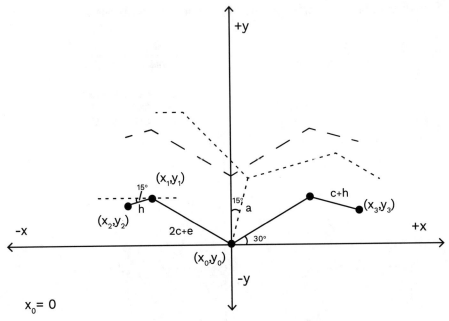

$x_0 = 0$

$y_0 = 0$

$x_1 = (2c+e) \times \cos(30°)$

$y_1 = (2c+e) \times \sin(30°)$

$x_2 = x_1 + h \times \cos(15°)$

$y_2 = x_1 - h \times \sin(15°)$

$x_3 = x_1 + (c+h) \times \cos(15°)$

$y_3 = x_1 - (c+h) \times \sin(15°)$

Generating the Motif

```
//scale factor
let a = 60;
let b, c, d, e, f, h;

function setup() {
    createCanvas(400, 400);
    angleMode(DEGREES);
    noLoop();
    noFill();
    b = (a * sin(120)) / sin(45);
    c = (a * sin(15)) / sin(45);
    d = (c * sin(90)) / sin(30);
    e = (d * sin(60)) / sin(90);
    f = (e + c) * sin(45);
    h = ((b + f) * sin(30)) / sin(60) - f;
}

function draw() {
    push();
        translate(width * 0.5, height * 0.5);
        let x0, y0, x1, y1, x2, y2, x3, y3;
        x0 = 0;
        y0 = 0;
        x1 = (2 * c + e) * cos(30);
        y1 = (2 * c + e) * sin(30);
        x2 = x1 + h * cos(15);
        y2 = y1 - h * sin(15);
        x3 = x1 + (c + h) * cos(15);
        y3 = y1 - (c + h) * sin(15);

        for(let i = 0; i<12; i++){
            push();
                rotate(45);
                rotate(i * 30);
                translate(0, -a);
                if (i % 2 == 0) scale(-1, 1);
                //every two step mirrored on the y-axis
                beginShape();
                vertex(x0, -y0);
                vertex(x1, -y1);
                vertex(x3, -y3);
                endShape();
```

Generating the Motif

```
            beginShape();
            vertex(-x0, -y0);
            vertex(-x1, -y1);
            vertex(-x2, -y2);
            endShape();
        pop();
    }
  pop();
}
```

Analyzing the Tessellation

Step 3: We need to calculate the dx, dy, and doff values in the placement.

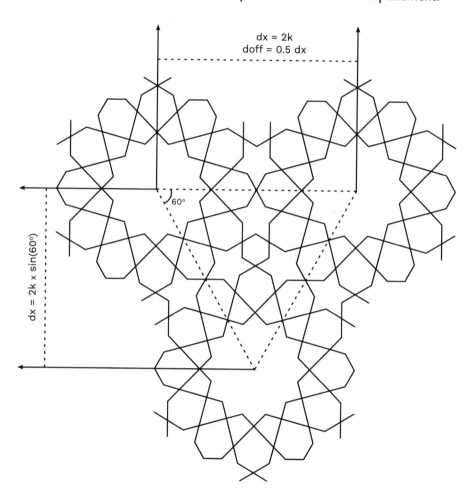

Tessellation Code

```
//Motif class
class Motif {
    constructor(a) {
        this.a = a;
    }
    display() {
        let b = (a * sin(120)) / sin(45);
        let c = (a * sin(15)) / sin(45);
        let d = (c * sin(90)) / sin(30);
        let e = (d * sin(60)) / sin(90);
        let f = (e + c) * sin(45);
        let h = ((b + f) * sin(30)) / sin(60) - f;

        let x0, y0, x1, y1, x2, y2, x3, y3;
        x0 = 0;
        y0 = 0;
        x1 = (2 * c + e) * cos(30);
        y1 = (2 * c + e) * sin(30);
        x2 = x1 + h * cos(15);
        y2 = y1 - h * sin(15);
        x3 = x1 + (c + h - 1) * cos(15);
        y3 = y1 - (c + h - 1) * sin(15);

        for(let i = 0; i<12; i++){
            push();
                rotate(45);
                rotate(i * 30);
                translate(0, -a);
                //every two step mirrored on the y-axis
                if (i % 2 == 0) scale(-1, 1);
                beginShape();
                vertex(-x0, -y0);
                vertex(-x1, -y1);
                vertex(-x2, -y2);
                endShape();
                beginShape();
                vertex(x0, -y0);
                vertex(x1, -y1);
                vertex(x3, -y3);
                endShape();
            pop();
        }
    }
}
```

Tessellation Code

```
//scale factor
let a = 32;

let motif = new Motif(a);

let nRow;
let nCol;
let dx, dy;

function setup() {
    createCanvas(800, 800);
    angleMode(DEGREES);
    noFill();
    noLoop();
    let b = (a * sin(120)) / sin(45);
    let c = (a * sin(15)) / sin(45);
    let d = (c * sin(90)) / sin(30);
    let e = (d * sin(60)) / sin(90);
    let f = (e + c) * sin(45);
    let h = ((b + f) * sin(30)) / sin(60) - f;
    let k = (b + f) / sin(60);

    dx = 2 * k;
    dy = 2 * k * sin(60);

    //approximate the nRow and nCol values
    nRow = 1 + ceil(height / dy);
    nCol = 1 + ceil(width / dx);
}

function draw() {
    for (let r = 0; r < nRow; r++) {
        for (let c = 0; c < nCol; c++) {
            push();
                if (r % 2 == 1) {
                    translate(dx * 0.5, 0);
                }
                translate(dx * c, dy * r);
                motif.display();
            pop();
        }
    }
}
```

I

Index

S. Artut, *Geometric Patterns with Creative Coding*,
https://doi.org/10.1007/978-1-4842-9389-8

Printed in the United States
by Baker & Taylor Publisher Services